ROAD TO THE SEA

By Florence Dorsey

MASTER OF THE MISSISSIPPI:
Henry Shreve and the Conquest of the Mississippi

ROAD TO THE SEA:
The Story of James B. Eads and the Mississippi River

JAMES B. EADS
oil painting
owned by Mr. and Mrs. James Eads Switzer

THE STORY OF JAMES B. EADS

ROAD TO THE SEA

AND THE MISSISSIPPI RIVER

by Florence Dorsey

A FIREBIRD PRESS BOOK

PELICAN PUBLISHING COMPANY
Gretna 1998

Manufactured in the United States of America
Published by Pelican Publishing Company, Inc.
1000 Burmaster Street, Gretna, Louisiana 70053

To my daughter
JANE WELCH

FOREWORD

This biography is compiled from the voluminous writings and utterances of James B. Eads, from government documents relating to his projects and the many controversies over them, from county and national records, magazine and newspaper accounts of his professional and social activities, books by many authors concerning his work and his times, and from information supplied by members or connections of his family. Outstanding among the books studied in preparation of *Road to the Sea* are Calvin Milton Woodward's *History of the St. Louis Bridge* and Elmer L. Corthell's *History of the Jetties at the Mouth of the Mississippi*.

I am indebted to many persons for material and research with which to supplement my own five years of research in the life and works of the great American engineer. Prominent among them are the following:

James Eads Switzer, of Long Island (grandson of James B. Eads), and his wife, Antoinette Hewitt Switzer, for permission to read and use some letters written by James Eads and his first wife, Martha, to each other, for having portraits of James and Martha Eads copied for my use; and for other courtesies.

Louis How (grandson of James B. Eads), for correspondence in locating certain Eads papers. His excellent pocket edition biography of his grandfather was used frequently for reference.

Wallace McHenry, of St. Louis (grandson of Eunice Hagerman Eads, the second wife of James B. Eads), for a fund of data about the home and social life of Eads, for light on the travels of the Eads family, and for accounts and descriptions of the five Eads daughters.

A. D. Stevens, of St. Louis (grandson of Sue Dillon Stevens, who was a sister of Martha Nash Dillon, first wife of James B. Eads), for many items about the family, home life, social activities, and character of James B. Eads.

Tom K. Smith, Chairman of the Board of the Boatmen's National Bank of St. Louis, and Mrs. Tom K. Smith (granddaughter of Erastus Wells, who lent his effort and influence to set Eads's various projects on their way), for valuable material. There were bits of information, too, from the historic bank's River Museum.

Frederick C. Ault, Librarian of the Municipal Reference Library, St. Louis, for information and solicited advice.

Dr. William G. Swekosky, of St. Louis, for his practiced research among records, his tireless hunts for obscure historical data, his personal investigation of St. Louis sites and reconstruction of old backgrounds, for photographs and charts, and thoughtful replies to innumerable questions with which I bombarded him for many months—unstinting and indispensable aid.

Marjorie Douglas, Curator, Brenda R. Gieseker, Librarian, and Charles van Ravenswaay, Director, of the Missouri Historical Society, St. Louis, for research, cataloguing, copying, advising, and other help—a year of it.

Marie H. Roberts, Reference Librarian of Washington University, St. Louis, for research among and reporting upon the Eads Papers in the Missouri Historical Society, St. Louis.

Rt. Rev. Msgr. John P. Cody, Chancellor of the Catholic Archdiocese of St. Louis, for directing searches into church records and supplying copies of the findings.

Charles H. Compton, Librarian, Mildred Boatman, Chief of the Reference Division, and Edith Varney, Technology Division, of the St. Louis Public Library, for help and courtesies covering several years.

Clarence E. Miller, Librarian of the St. Louis Mercantile Library, for answers to queries.

Francis C. Burgess, Superintendent of Bellefontaine Cemetery, St. Louis, for many delvings into records.

P. J. Watson, Jr., President of the Terminal Association of St. Louis, for material.

Captain Donald T. Wright, publisher and editor of the *Waterways Journal,* St. Louis, and his business manager, Mr. Swift, for research.

Floyd C. Shoemaker, Secretary of the Missouri State Historical Society, Columbia, Mo., for data and cataloguing.

Henry C. Strippel, Head of the Genealogy and Local History Division, New York Public Library, and his assistants.

F. Ivor D. Avellino and Sylvester L. Vigilante, of the American History Division, New York Public Library.

Staff of the Science and Technology Division of the New York Public Library.

Staff of the Economics Division of the New York Public Library.

Staff of the Newspaper Division of the New York Public Library.

Raymond N. Brown, Engineering Societies Library, New York.

Mrs. E. D. Friedrichs, Archivist, City of New Orleans, and her assistant, Marie Clark, for research and voluminous copying from New Orleans newspapers.

George King Logan, Acting Librarian, New Orleans Public Library.

Mrs. J. Frank Gordon, Le Claire, Iowa, for research, and for the loan of old newspapers from her historical collection. Most of the Le Claire background was provided by Mrs. Gordon.

Edna Giesler, Librarian, Public Library of Davenport, Iowa, and her aide, Jean Wagner, for research and copying.

Luther Harris Evans, Librarian, Library of Congress, for special cataloguing of sources of material.

Philip M. Hamer, Records Control Officer, National Archives, Washington, for special cataloguing of sources of material.

Henry S. Parsons, Chief of Serials Division, Library of Congress, for having research done and photostatic copies made.

Jane Welch, New York, for research in Congressional records and documents, and in foreign journals.

CONTENTS

Chapter		Page
I	James Eads Discovers the Mississippi	1
II	Diving for Sunken Treasure	10
III	The Bell-Boat Fleet	26
IV	Country Gentleman	39
V	The First Ironclad Gunboats	49
VI	The River Monitors	77
VII	Spanning the Big River	96
VIII	Below the River Bottom	107
IX	The Long Strides	129
X	The Great Bridge	155
XI	The Gulf Bar	166
XII	Willow Walls	181
XIII	The Proud Little Pass	201
XIV	The Tragic Isthmus	218
XV	A Ship-Railway	238
XVI	Deep Channel	260
XVII	The Long Fight	270
XVIII	The Tired Warrior	282
XIX	The River Rolls on	295
	Notes	307
	Condensed Bibliography	333

CHAPTER I

JAMES EADS DISCOVERS THE MISSISSIPPI

L ATE in the summer of 1833, a steamboat moved down the Ohio River on its way from Louisville to St. Louis. Tired-foliaged trees on the high banks cast their reflections far over the water, leaving an irregular middle strip that mirrored only the sky, the shore shadows devoured the ripples that hurried away in each direction from the prow. An occasional barge or keelboat, the crew bending to oar or pole, drew aside to let the arrogant steamboat pass. Flatboats steered by long sweeps pulled laboriously out of the way. The hills ahead opened slowly, widening the river for it, the hills behind closed silently after it. White roustabouts idled on the lower deck beyond the engines, lolling on the freight and baggage. Negroes trotted errands through the long saloon cabin above. On the upper deck lounged the passengers. Men in long-skirted coats, tight pantaloons and tall beaver hats talked about states' rights, tariffs, horse racing, gamecocks and, most of all, about steamboats and the rivers. It gave a man a sense of power to live in an age of fast, luxurious travel. Bets were tossed recklessly on which of the elegant boats was the speediest, carried the most passengers and freight, had the best-stocked bar or served the greatest variety of desserts at a single meal. A few women, gathered in sheltered places out of the wind, their handsome shawls falling back from full skirts, snug bodices and leg-o'-mutton sleeves, chatted about styles and the theater, their long narrow bonnets shifting a pattern of varicolored cylinders as they turned from one to another. Louisville was fashionable and worldly, with the

French modes, silks and cosmetics of New Orleans, and the best touring drama of the East seeking it out.

Somewhat apart sat Ann Eads, with her two daughters, Eliza Ann and Genevieve, sixteen and fifteen years old. Ann was a neatly dressed woman, retiring yet arresting. Her finely chiseled face, strong but wistful, seemed washed from within by a white light of patience, her shoulders bent forward as if to accept whatever burden might be laid upon them next. She was very tired now, for she had been packing to move—her household goods and trunks were stowed on the lower deck. Her husband, Colonel Thomas Clark Eads, descended from a substantial Maryland family, had not prospered in Louisville, nor in other towns farther back on the Ohio, and had drifted from port to port with his wife and children, beginning each venture in his graceful, lordly manner only to see it dwindle to failure. He was going to make another start, in far-off St. Louis, when he had collected stock for a general store, and was sending Ann on ahead to prepare a home for him.[1]

Much of the time Ann was preoccupied with listening for distant sounds—somewhere, ranging the boat, was her thirteen-year-old son, James. A patter of running feet, and she would lean back. A thin, wiry boy, with light-brown hair tousled in the wind, gray eyes dancing, would slip to her side breathlessly with a tale of wonder about the boat's engines or passing craft. His manner was gentle, but his energy was appalling and his pranks legion. One never knew what he would be up to next, and there was a while in Louisville when his father hired a sturdy playmate to take the brunt of his mischief.[2] James Buchanan Eads, born in Lawrenceburg, Indiana, May 23, 1820, and named for his mother's cousin, James Buchanan, a young Pennsylvania lawyer who had just been elected to Congress,[3] was destined to be up to something unusual all his life. As one of the greatest engineers of all times[4] he would shape his country's progress and play a strong part in the international drama, a compelling figure, adored and hated, honored, neglected, triumphant, tragic.

[1] Numbers refer to notes, all to be found at end of the book.

He was going to build a steamboat when he got to St. Louis, he confided to his mother now. It would be much harder to make than the puffing steam engine he had put together three years ago in his workshop after they had moved from Cincinnati. As for the sawmills and fire engines he had made, they were nothing. This was going to be a real boat and run by its own steam.

Off he padded again on a round of investigation that took him from pilothouse to paddle wheel. Then for a while he forgot the boat details, the shore hills had caught his eye, marching away inland to slip at last behind a turn of the bank, leaving a low flatland. Willows bent over the water, dabbling their fingers in it, they moved back from an army of reeds that climbed ashore, they rallied again with a canebrake pushing behind them. A rim of white curved ahead with the horizon. There was a scraping of chairs on the deck. "The Mississippi, the Mississippi," everyone was saying. James hurried to the bow.

Others crowded behind him. The white rim came nearer, it spread wider, tipped up by the distant Missouri shore until it stood almost on edge. Now it was flattening out, getting longer, winding narrowly out of the northern sky, stretching itself broadly here and swaying on southward to meet the sky again. Its waters lashed nearer, rougher, they caught at the boat greedily and fell away from its prow in a boiling double spray. James pressed against the rail, his hands tightened on it—nothing so exciting had ever happened to him before. If you look a long time at the water you felt a very part of it, galloping, slashing, romping away with it. The Mississippi was different from anything people had said about it, more real, more alive. It could be happy, angry, proud or reckless, as though it had thoughts and feelings. It must be the most wonderful river in the world—his river.

In a boy's realm everything exists for the boy, but this was always to be James Eads's river. He was to understand the vast willful stream as no one ever had before, sail its surface, walk its bed with the flood surging over him, build it a navy, span it with steel, deepen its channel, shift its mad torrent at will, work for it, fight for it, and die in an exhausting effort to open its rich commerce to the Pacific.

The gentlemen behind him were exclaiming over the change in the river—it used to be cluttered for twelve hundred miles with groves of dead oaks, cottonwoods and sycamores that had been toppled from high banks by undermining floods and transplanted in the soft bed by the current. For centuries Indian canoes had picked their way in and out of these half-submerged forests and around tangled masses of timbers that stacked against shore points and islands, the white men's boats had not dodged about so easily, and many a hull was ripped by branches and log ends reaching under the opaque water. But lately Henry Shreve had invented a strange twin-hulled snag boat that had pulled up dead trees and torn out matted drift logs until the Mississippi spread away, clean and open, like a choppy inland sea. And now more steamboats were sailing up the rough stretch from the Ohio mouth to the once nearly isolated St. Louis.

The few days that the steamboat labored on northward against the heavy Mississippi current were filled with enthralling sights and sounds for a boy. The river lapped teasingly at the Illinois prairie whose tall sear grass billowed in the wind as far as he could see, it lashed angrily at the high Missouri bank that swallowed the afternoon sun, it laughed at the foot of a natural stone tower rising sheer from the water, it whispered past rock caves where pirates used to lurk—maybe even Mike Fink himself. It slapped at islands noisy with migrating birds, it rocked flatboats gleefully, setting their loads of cattle to lowing and dogs to barking. It crept stealthily at night when the landing torches threw an orange glow over the face of the bluff and misshapen shadows bobbed up and down in it as roustabouts ran back and forth with burdens.

One night James's mother made ready for the journey's end—the boat would arrive at St. Louis early the next morning, September 6. She hurried him to bed. It seemed hardly a minute before she was shaking him awake. Something had happened, he could hear men running, alarm bells clanging. A shout of "Fire,

fire," rose above the din. He had only begun dressing when his mother herded him and his sisters out the door and into a milling line of half-clad passengers pushing toward the deck. The air was full of cried names as people got lost from each other, gusts of smoke, pungent with the smell of burning painted wood, screening them apart. A shift of breeze cleared the air for a moment. Yonder in the dawnlight lay St. Louis on the edge of a terraced bluff. As the boat panted to reach it, men with black-smeared faces fought the flames that licked at the backs of those huddled forward of the cabin. There was a sudden surge of the crowd, a relieved babble, but a boy could not see what was going on. Then, barefoot and coatless, James stood shivering on a low sandy bank at the foot of the bluff.[5]

His mother trembled as she shielded him from the chill wind. Other passengers thronged past them, climbed a steep street and vanished, and still the little group stood. They had no place to go, they were destitute. Everything they possessed, their furniture and clothes, fed the flames that leaped to hide in the smoke circling in the morning mist. Men pulled fire pumps by, shouts greeting them faded into the crackle of wood and hiss of steam. Somewhere a church bell was ringing. People seemed to rise from the sand to see the boat burn—they had been on their way to the little cathedral yonder, some were telling each other. Several ladies stopped beside his mother, questioned her, looked at the charring boat. Then they were all climbing the steep street.

After this everything was like a rapidly changing dream . . . shelter in the home of one of their kind new friends . . . his mother moving to a place she had rented, the upstairs of a house facing toward the river, on a street that ran along the top of the bluff a short distance from its edge—North Main Street, they called it . . . a setting to rights of odd pieces of furniture . . . his mother bending over a cookstove, his sisters carrying heaped dishes into the dining room to pass down the long table to businessmen who came there for their meals.[6]

When he was not needed to help about the house, James hunted a job so that he might do his part until his father came to take care of them. But it seemed that nobody wanted to hire a boy who did not look strong. Trudging across the town one discouraging day his eye fell upon a wagon heaped with apples. It was an inspiration. Everybody liked a good apple to munch—he bought a basketful and began to peddle them.[7] After this he was part of the kaleidoscopic scene that shifted its colorful bits on the busy streets, a thin, pale boy, bunted by the river wind, selling apples from a basket on his arm and looking a little wistfully at children on their way to school.

It was exciting, though, having a business of one's own, and while he was attending to it he could do some exploring. St. Louis was a big town, it reached a long distance on the bluff edges, and had six thousand people. Elegant gentlemen in tight-waisted coats and tall hats, farmers in homespun, backwoodsmen in leather pantaloons and jackets, boatmen, Negroes and even Indians jostled each other on the narrow sidewalks. A new guidebook described Missouri as "bounded north by the Sioux district, east by the Mississippi . . . south by Arkansas Territory, and west by the Osage and Sioux districts." There was a good deal of talk about moving all the Indians out of the state to public land farther west. Handsome sleek carriages wove in and out among the oxcarts on the long streets that paralleled the river, some of them pulled up to hotels so fine that it awed one just to peep through a window into their lobbies. There were plain hotels, too, with rough board floors and homemade furniture in their parlors—James had seen the patrons lined up outside the back door in the morning, waiting a chance to douse their faces at a wash basin on a high bench.[8]

Sometimes he stopped to watch the stagecoaches rattle in from faraway Vincennes or Peoria. The passengers looked tired and exasperated, having had to lighten a stalled coach by floundering afoot through a bad piece of road. Or they were drenched after riding unhitched horses over a swollen stream to take their places again in the soaked vehicle after it had been dragged across and bailed out. Travel by land was terrible, everybody said, even on a railroad in the East where one got jerked and jolted, singed by flying sparks or deafened by the snorts and clatter of the locomotive. The magic way to travel, James knew, was by the rivers. He liked to perch on a bluff edge where he could look at the Mississippi. Its waters came endlessly out of the north to lap and swirl forever southward.[9] All other moving things stopped and started, unloaded and filled, but the river was ceaseless. While people worked or while they slept, the river flowed on. Nothing could be more constant. Ferryboats crossed it with passengers and baggage, barges sidled past in it, the boatmen jabbing the water with long poles, flatboats drifted on its ripples, twisting with the current. And often there was a steamboat cleaving it into spray. Three years ago only eight steamboats had come to St. Louis in a twelvemonth, it was said—last year there were eighty arrivals![10] When James, his basket empty, started home at dusk the patter of ox-hoofs and creak of cart wheels had nearly died out, minstrelsy was drifting from poolroom and tavern past the faint twinkle of oil lamps.

Indian summer passed, the days grew wet and windy. People hurried past on the streets without a glance at the basket of apples held out temptingly, and there were too few coins at the end of a day. James wanted a job, a real job with steady pay, and he probably said so wherever he thought it might count. One of his mother's boarders, Mr. Barrett Williams, who had a drygoods business only two blocks away on the corner of Main and Locust streets, suddenly found that he needed an extra hand to run errands and do odd chores and that James was just the boy for it. The new lad-of-all-work at Williams and Duhring's delivered parcels, carried out ashes, fetched in coal, tidied about with a will, and proudly brought home his pay of three dollars at the end of the week.[11]

In his leisure time he collected materials for the steamboat he was planning to build, and he invented an unusual vehicle, the only one of its kind. One evening he brought his completed model into the living room and set it down on the floor. It ran violently here and there, startling his mother and some callers who had dropped in. The young inventor's explanation of the motive power did not put the others at ease—it was only a big rat pulling the cart about. He had fastened the rat by its tail on a treadwheel which gave motion to the cart wheels, and a hole had been left open in the narrow cover just out of reach of the rodent, showing a way of escape. The lusty kicking to reach the hole whirled the treadwheel.[12]

Only his employer, Barrett Williams, fully appreciated James's invention. He found the boy remarkably original, his mind reaching out hungrily for expression—it was a pity that his school days had been cut short. Colonel Eads, a fine-looking man of commanding presence, had opened his business in St. Louis and appeared to be doing well, but James seemed to feel that he must keep on at work, perhaps remembering vividly the collapse of other of his father's ventures. In his modest quarters over the store, Barrett Williams had a cherished collection of books,[13] many of them on scientific works, and he decided to turn the ambitious young mechanic loose among them.

Books, a roomful of books! the chore boy marveled as he was led into the sanctum. Books about machinery, ores, atoms and stars—about everything. A new world spread away, and there were no right words with which to thank the kind man who had swung back the door to it. After this the evening tasks made record time, dust flew from the broom, kindling and coal for the morning fire appeared in a twinkling, supper was put away with dispatch, and James climbed the stairs to the book-lined room. By the light of an oil lamp he adventured into exciting realms of physics, mathematics, mechanics, history, poetry, the classics.

This round of work, study and play went on without recorded incident for three years, except for the death of Genevieve,[14] the younger of James's sisters, whose frail body had long been shaken by a consumptive cough. Then, early in 1837, the contented regime was threatened—Colonel Eads was moving on to pastures that looked greener to him. On a trip up the river to sell some goods he had fallen in with Mr. Eleazar Parkhurst, a one-time shoe merchant of St. Louis, and was joining him in the development of an Iowa town, to be called Parkhurst, at the head of the upper Mississippi rapids, about fifteen miles above Davenport.[15] James felt pulled in two directions. He wanted to be with his parents, but he had a good job—he had been promoted to salesman long ago—the library above the store, where new books were always appearing just as they were needed, held him in strong, pleasant bonds. And there was the steamboat, longer than himself, that he was building in odd hours. He would stay.

After the wrench of decision, life flowed on placidly again. The evenings in the book room grew longer. The jingle of harness, the thrum of a guitar, the bellow of tavern roistering, would fade into stillness. Across a long quiet might sound the bawling of a late reveler or a distant cockcrow. James Eads, of little schooling, his shoulders hunched over a table, his shock of light-brown hair and his earnest young face burnished by the lamp flame, was laying the foundation for a knowledge of physical laws, and of man's power to wield them, with which he was to hew out such daring paths for engineers as to set two continents marveling.

CHAPTER II

DIVING FOR SUNKEN TREASURE

FIVE years of running errands, doing chores, measuring cloth, tying packages and poring over books had slipped by since James Eads had laid aside his apple basket—it was now the autumn of 1838. The world about him had changed swiftly. St. Louis, a door to the four quarters of the nation, stretched several miles along the high terraced riverbank and straggled back inland to Seventh Street, a new section "almost too far out for dwelling."

The store of Williams and Duhring's looked out upon the valley's busiest land and water roads. Behind the counter, James Eads, eighteen years old, slight, with straight sensitive mouth and imaginative gray eyes, found time between sales to take in the scene that wove in patches past the door and windows. Not far down the waterfront was the great Market, and all about it swarmed carts, wagons, carriages, and customers afoot. Men of every rank bought and sold, haggled and bargained, those of the gentry picking fastidiously from stall to stall, Negro servitors following them with baskets. The streets above the Market streamed with vehicles. Long dusty caravans wound in from the Santa Fe Trail to the shouting of drivers and cracking of whips, each wagon drawn by six or eight oxen, bringing a wealth of Mexican "hard money." Fresh caravans heaped with merchandise, principally cotton print and baubles, plodded westward to set out across the plains.[1]

Beyond the Market and the busy waterfront reared the triple-storied steamboats, their white filigree woodwork glinting in the sun, their twin smokestacks a black colonnade against water and sky. Often there was a mile's length of steamers at the wharves—the steam tonnage of the valley rivers was double the entire steam tonnage of England, and the best of it in the St. Louis trade. The waterside was piled with barrels and crates, drays loaded and unloaded. Gentlemen in walking coats, long pantaloons pulled over snug Wellington boots, and tall hats accenting the fashionable muttonchop whiskers, went aboard the steamers, or ashore, their baggage trundled after them by slaves or wharf hands. Women, their pointed Cashmere shawls falling over full skirts, found their way daintily in and out of the carts and stacked freight.[2]

Although he might be waiting upon a customer James could hear the gun signals, the clang of bells, the song of cabin hands, the pant of engines and slosh of paddle wheels—a steamboat swinging its blunt nose to shore or starting away to wind with the broad river between bluffs and prairie, canebrake and cypress swamp. A longing for the river, to be back on it, to feel the motion of it under his feet, was always with him. But there was not much time to indulge a fancy, for business was brisk at Williams and Duhring's, it was good everywhere in this thriving port. The financial depression that had been spreading ruin over most of the country for a year had hardly touched St. Louis, fended as it was by Mexican gold and silver coming with every caravan from the southwest.

Not even the advent of a bank, recklessly established by a group of busybody St. Louis gentlemen, had yet disturbed the town's prosperity, but men whose boots tracked the sand of the waterfront and mud of farms over the store floor still discussed its threat, their voices rising indignantly. No institution, they declared, was as dangerous and reprehensible as a bank, which only "intoxicated business, then left it prostrate." There was usually some excited debate going on in the store and the young salesman could keep an ear cocked to it. One of the liveliest had been over a proposal to secure a railroad through the state, a plan that looked feasible enough if horses were used in occasional hard pulls to aid the locomotive. No topic had ever brought forth more contemptuous snorts around the store grate. Men along the river knuckling to the railroad zanies! Rashness was in the very air![3]

When the store closed after dusk James could stretch his legs awhile before settling down to his books. Usually it was the river that drew his exploring feet. All day its gray or sunlit ripples had beckoned, now it lay secretive under the prying stars, tossing aside gold sequins wavered across it by the port torches and creeping stealthily into blackness. A steamboat riding off with it into the dark became a jewel-crusted bauble hung in the night.

On Sundays, when men by the scores were deserting the wide-open taverns to clatter by cart or horseback to the race track outside the town, James Eads would trudge to Chouteau Pond, a beautiful lake west of town, lugging the steamboat he had made. It was an armful, six feet long, complete in machinery, multiple decks and smokestacks. Picnic parties dotted the wooded shores, tiny sail boats drifted about on the water. Presently James would have his steamer chugging over the lake, smoke spouting from its chimneys, ripples widening from its path. A rim of spectators would gather on the bank to watch it.[4]

Midwinter of 1838-39 was bleak and gray. The pond was stilled, the Mississippi a broad ribbon of ice that pressed the steamboats motionless against the bank. The very silence of the river was awing. When the early spring thaw set in and the imprisoned water burst its ice shell, flinging heavy fragments downstream and lunging after them, James could feel his own spirit racing along with it, boisterous, free. He had had enough of land.

On a March morning of 1839, he walked briskly along the waterfront, his scant baggage bumping against his legs, a blustery wind whisking his circular cape about his steamboat which he carried under one arm. There had not been many persons for him to tell good-bye: his employer, Barrett Williams, who had always understood how a boy felt about things, the kindly priest at the little Cathedral Church of St. Louis de France, his cousins,

Susan and Martha Dillon, out at Rose Hill. Martha, sixteen, slim and dainty, her brown hair catching the light, wide blue eyes full of questions, understood too. He was going to make his home with his parents in the new village up the river, Parkhurst, he had confided to her, but he would stop in St. Louis to see her whenever he could.[5]

He turned up the gangplank of the *Knickerbocker,* a steamboat in the northern lead trade, and reported as the new second clerk. The boat backed away from the bank amidst the usual shouts and bell signals, and headed upstream, the engines laboring against the current until the heavy inpour of the Missouri was passed. The limestone bluffs that had reared on the west subsided to a flatland, the low Illinois shore grew higher until it faced the river with sheer cliffs—the mile-wide stream, only a thin remnant of the mighty torrent that had, millenniums ago, surged down the valley and cut a deep way through ancient coral beds, now wandered from one prehistoric steep bank to the other, leaving a bottomland that reached away to a deserted bluff standing in the fields of a dry countryside.

The job of a second clerk—"mud clerk," he was called because it fell to him to wade the slush of unpaved waterfronts, accepting or releasing freight—was not an easy one, James found. He stood watch with the mate from noon until supper, then turned in while the captain and head clerk took over. At midnight he pulled himself from his berth, collected freight bills, bargained with the rapacious woodyard "pirates" for fuel, kept one eye on the all-night gamblers aboard, and the other on the riffraff deck crew to make sure that they did not tap the barrels of whisky always on hand. After he had crawled wearily back to his bunk he was likely to be routed out again if the head clerk felt the weight of his duties too keenly.[6]

But James was barely mindful of hardships, he could hear the splash of the paddle wheel, feel the throb of engines under his feet, the fresh wind on his face. He prowled the boat, peering at every part of it, he spent hours of his scant leisure wondering at the river and its moods. It dug its channel here and built a shoal yonder, lashed savagely at stone bluffs, idled around a sandy point, murmured and sang, then raged in frenzy for miles in the Des Moines rapids. Its fury spent, it rippled serenely for a piece before it lunged in and out of the stone ledges of the long Rock River rapids. The steamboat passed the infant town of Parkhurst on the west, meandered on to Galena near the upper tip of Illinois, took on part of its cargo of lead, moved up to Dubuque, an Iowa mining town, and groaned heavily downstream, drifting over the stone ledges in a spring freshet.

Whenever the boat stopped near enough Parkhurst, James made a hasty visit to his parents. Colonel Eads had built, as a much more appropriate setting for himself than the modest St. Louis boarding flat, a large frame hostelry, the "Berlin," considered thereabouts "a wonder in magnitude." As a dignified and kindly host, always dressed "in regulation black with swallow-tail coat and stove-pipe hat," he had drawn to him the social life of the steadily settling wilderness. Balls, wedding receptions and church services were held in Hotel Berlin, and newcomers to the section liked to snuggle their cottages near it for human warmth—across the street Isaac Cody, yet to be the father of "Buffalo Bill," raised his house. The Berlin was a gay spot of sophistication, its night lights twinkling against forested hills. It was a political center where issues great and small were argued. And it was a place to which the troubled came for comfort. Ann Eads—Nancy, her husband called her—put aside her grief over the death of her older daughter, Eliza Ann, from the same dread malady that had carried the younger Genevieve away, and gave her motherly counsel and sympathy without stint. His stay here was like a swift panorama to James, he marveled at his impressive father, made a quiet ado over his mother, and hurried back to his duties on the lead boat.[7]

Before the season was over, the *Knickerbocker,* rounding from the Mississippi into the Ohio, was ripped open by a snag, and for the second time James Eads escaped in the early morning, half dressed, from a doomed vessel. With passengers and the rest of the crew he was picked up by a passing flotilla of flatboats just as the heavily laden steamer tilted down, the bellow of a few cattle aboard choked in a hungry gurgle of water.[8]

The *Knickerbocker* was gone. It gave a very young man a strangely lost feeling to have the deck of his boat sink from beneath his feet. Rivermen spoke of the "death" of a boat when it vanished in the river, but perhaps many of the craft that lay in the long winding graveyard of water and sand were not dead, but only lying there at the bottom guarding their secrets of cargo. Berthed, directly, on another steamboat, James Eads fell to wondering about the freight that lay hidden in drowned hulls all up and down the rivers. He toyed with the notion of treasure hunting in the submerged wrecks—if only he were a diver, if he had a boat of his own that he could stop wherever he liked, take off from its deck, sink down in the opaque stream and grope about on its bed, what might he not find! A diving-bell boat! He meant to get one.

There was really no such thing, he was told when he made inquiries, except a few makeshift contraptions. No good diving-bell boats? Then there should be. Whenever there was an hour for dreaming the tired mud clerk pictured, in fancy, his bell boat. Submarine, he called it. It grew sturdier, more impressive, all sorts of devices taking their places on it. By the end of his third year on the river its designer quit his job to go about getting it built.

And so it was that the astonished heads of a St. Louis boat building firm, Messrs. Calvin Case and William Nelson, were confronted in their office one morning by a young man who, with lighted eyes and a pleasant, determined voice, offered to take them into the salvaging business with him if they would construct the new vessel he had devised. Their amazement shifted to the sketch he spread before them. It showed a stout twin-hulled boat with a sparse grove of derricks and pumps rising from its deck. There was an eager explanation of how the craft was worked and what it could do. When the conference was over, the two boatbuilders found themselves partners of the twenty-two-year-old inventor.[9]

With considerable pride James Eads kept close track of the progress of his bell boat, part of which was being constructed at Paducah, Kentucky. It would do everything but the diving, and he would do that. But before the Submarine was completed a tempting contract in salvaging—or "wrecking," as it was known along the rivers—was offered his new company, the recovery of a hundred tons of lead from a barge that had sunk in the Mississippi rapids not far from Keokuk, Iowa. He had a scheme for raising this lead, James confided to his partners, and it would be a good opportunity for him to get a lesson in diving. He engaged a professional Great Lakes diver from Chicago, rigged up a barge with a derrick and air pump, had it pulled up the river and anchored over the wreck. When the diver arrived all was ready for him. He got into his armor and tried to descend to the wreck, but the staving current pushed him aside. A rope was then fastened to his belt and to the bow of the barge, but still the savage water threw him about. He clambered back on board. This was not lake diving and he had had enough of it. He was quitting.

James had to think fast, he could not let his first wrecking effort fail. He persuaded the diver to stay and make another attempt in a sort of diving bell which he would have on hand directly, and set out for Keokuk. There he bought a forty-gallon whisky barrel, brought it back to the barge, fastened several pigs of lead to the lower open end to sink it and a block and tackle to the closed upper end, strung a rope from the boat's bow to hold the barrel against the current, and the improvised diving bell was complete. The diver looked at the thing sourly and refused to enter it.

He would have to turn diver himself at once, James saw, deciding then and there that he could never ask a man to take a risk that he would not dare himself. Borrowing the diving harness, he sat down on a strap that he had placed across the lower end of the barrel, and ordered his crew to let out the derrick ropes and swing him down to the wreck. With the current beating against his bell he groped about, slipped a loop end of cable around a lead pig and jerked a signal cord. After the pig had been drawn up and released, he pulled the cable back for another load. When he had sent up all the lead lying under the bell, he gave a different signal, a guy line moved his bell a trifle upstream and the work went on.

After a while the water pressure began to tell on the novice and he signaled to be brought up to rest. At once calamity turned loose. The diving bell had been moved so well forward as the work progressed that the derrick, not properly stayed, was toppled over by the pull on it. The crewmen grabbed the block and tackle and managed to haul the barrel up to the water surface, but could get it no farther. The air pump was still working, and the amateur diver waited and wondered. A hand groped under the rim of the barrel, he grasped it, crawled down from his seat and was drawn up. After he had had the derrick raised and made secure he was lowered again to the wreck. He had made several trips down and recovered a good part of the lead when the hired diver volunteered to finish the job.[10]

Not long after this, James Eads brought his Submarine out on the river. He sailed it to sunken boats in the Cumberland, the Tennessee, the Ohio and Missouri, but the Mississippi was his main hunting ground. Rivermen chuckled over the bristling craft —it would take a mud clerk to think up a thing like that! But the skipper of the Submarine, now known as Captain Eads because he was master of a craft, was not disturbed by random opinion. His adventure absorbed him. When his crew anchored the vessel over the reported location of a wreck with worth-while cargo, he would don his diving suit and go over the side in the bell. If the wreck was old, there might be a long search up and down hill of the river's rugged bottom before it was found, for the reports were often faulty, and the notionate river had a way of changing its course. He thus described, much later, his hunt for a drowned hull that had lain on the Mississippi bed for thirty years:

"Five miles below Cairo I searched for the wreck of the *Neptune* for more than sixty days, in a distance of three miles.... My boat was held by a long anchor line, and was swung from side to side of the channel, over a distance of five hundred feet, by side anchor lines, while I walked on the river bottom, under the bell, across the channel. The boat was then dropped twenty feet further downstream and I walked back again as she was hauled towards the other shore. In this way I walked on the bottom four hours, at least, every day (Sundays excepted) during that time." The *Neptune* was fifty-five feet below the surface when he found it, on a hard gravel bed. He brought up its cargo of lead, and sundry articles among which was a jar of butter "in a good state of preservation."[11]

Not only did the river change its course and baffle the searcher, but sometimes went to peculiar lengths to obliterate all signs of its prey. In after years James Eads often illustrated this with the story of salvaging the cargo of the steamboat *America,* sunk at Plum Point, about a hundred miles below the mouth of the Ohio. The soft bed there had sucked the hull down and down, the current had piled brush and sediment against its decks and chimneys until a substantial island was formed. Willows took over the new land, cottonwoods claimed it from them. When the wreck was twenty years old a farm was being tilled above it and cordwood from the trees sold to passing boats for fuel. Then the stream had edged westward until the island, lying at first near the Arkansas shore, was in mid-channel. Here floods began a capricious attack upon it, tearing strips from it in savage hunger, overturning trees, sweeping away buildings and denuding the decks of the wreck. In time, after more than two months of search, a bell boat sat where cows had once pastured, and a cargo now submerged forty feet, was rescued.[12]

The Submarine had not made its way on the streams very long before it was the talk of every landing. A new flight of treasure tales arose along the banks—nearly every villager or boat hand could tell a story, maybe an inherited one, of sunken wealth. Shippers and insurance companies appealed to Captain Eads to go hither and yon to save a costly cargo, offering as much as one-third or one-half of its worth. Ancient wrecks paid him even better, the salvager having sole right to freight submerged more than five years. There were few days when he did not go over the side of the Submarine to grope about for boat hulls.

He usually found the river bottom as engaging as the wreck. Its very wildness, its sudden bars, unexpected canyons and writhing turns, recorded in cryptic code the mild moments or frantic surges of the stream, the joy or anguish, the might and frustration of a giant. Bit by bit the diver captain was learning to read that code. He spoke of it sometimes as a record written by God Himself in the language of natural law.

Weeks of trudging the sucking silt of the river bed, of scrambling over tilted decks, pulling aside drift logs, removing sand, handling freight, in constant danger of getting his air line torn away by a snag or wrenched off as he stumbled into a sinkhole, and then a short rest ashore. It was good to get back to the village above the upper rapids to see his parents and have a taste of home life. Colonel Eads, busy with his land trading, had retired as host of Hotel Berlin, and built a house at the head of a beautiful narrow valley which Ann had named Glen Argyle for the seat of her family clans in Scotland. It was a roomy house, "prim and modest," its roof swept low to cover the long porch, two dormer windows peered through the trees down a lane.[13] Doubtless Ann had helped plan it, it was so homelike and gracious. To her "Argyle Cottage" was a haven after a long wandering. James always brought a bit of money to provide extra comforts for his mother or to help save one or another of his father's real estate ventures from disaster—the burden of these fruitless speculations and his sympathetic patience with them run in a nearly unbroken thread through James Eads's later letters.

But the high moments of James's stays ashore were when he landed in St. Louis and, dressed in his neat best, went to call at the Dillon home, ostensibly upon his first cousin, Eliza Eads, Colonel Dillon's young second wife, but really to see Martha, the colonel's twenty-year-old daughter. Martha, in a pretty slim-waisted dress, her brown hair smoothed down at the sides and caught up high on the back of her head in a roll from which curls dangled to her shoulders, her blue eyes discreet, expected him, for he always managed to get a message to her through her married sister, Sue. Cousin Eliza knowingly kept busy elsewhere, but her husband stayed on hand and invariably took over the occasion. The young couple sat a goodly distance from each other, getting in only a few words, addressing each other as "Mr. Eads" and "Miss Dillon," although they were related by blood ties, aside from the connection furnished by Cousin Eliza. Even their glances had to be cautious, for the host was never too engrossed in setting forth his various business successes to note all that was going on. Occasionally James and Martha met at the home of Sue and her husband, Dr. Charles Stevens, a young physician from the East. Now and then, in the summer of 1844, they went so far as to steal a few horseback rides together, and Martha had boldly given James permission to call her "cousin," which sent him back to the river warm with hope.[14]

Daringly he wrote to her, in September, working the letter around to a disclosure of his high ideals of married happiness. To his delight, Martha replied, her stilted note growing personal enough to compliment his fine penmanship. James rushed off another letter, modestly giving all credit for his good writing to his pen, which he enclosed as a gift. It was a plain little brown wooden holder with a sharp steel point, and it still lies in the hundred-year-old missive in which Martha immediately returned it in order to preserve her "self-respect," as she lacked her father's permission to receive gifts. "How long my father's roof may shelter me," she wrote with a Victorian sense of drama, "I know not, but while I remain within its precincts my line of conduct is determined upon." But she added provocatively, "Do not use this pen until you have occasion to write to her whom you have selected for your companion as you journey through this vale of life."[15]

This was a cue not to be slighted. James lost no time in dispatching a letter to amaze her with the fact that she was his choice. He pleaded his cause in a breathless succession of letters, arguing frankly in his own behalf. An honest man, he said, was "the noblest work of God," and he believed himself to be one. He had poured out his love for "The Sweetest Flower on Rose Hill" in many a finely written page when Martha at last unbent to thank him for "the tenderness of his language . . . the overflowings of a manly heart." This encouraged him to plead his case all the more fervently, even to swagger a little about himself and men in general. Back came a brief, spirited note, beginning aloofly: "Dear Sir: You wish me to accept as a fact that 'the Lords of creation' are stronger, their reasoning faculties far cleverer and sounder than ours?" and signed herself, "A motherless De Staël."

As soon as he could leave his work James started for St. Louis. All this was too oblique, too time-wasting, he liked things direct, to the point. Months had gone by, it was June of another year, 1845, and nothing settled. He met Martha at her sister Sue's home and begged her to marry him in August, only two months away. She countered by setting a far-off November date, provided that James could get her father's consent—it would not be kind or honorable to marry without it.[16]

Never lacking courage, James called at Rose Hill on a soft June evening. Colonel Dillon, as usual, took over the reins of conversation, and presently Martha slipped from the room to give her suitor an opportunity. When Dillon paused for breath in his description of a fire grate on Chouteau Avenue, James plunged into his mission. It bounced the older man from his chair, he walked over to the piano, gave a characteristic shake of the head and poured himself a drink of water. This had been a real shock to him, he had never suspected that anything of this sort was afoot. He looked the audacious fellow over: slender yet muscular, carrying himself with conscious pride, his face ruddy from outdoor work, his eyes deep and unflinching, mouth firm—an upstanding young man if there ever was one, but obscure, without handsome prospects. It would never do, the colonel decided. He had other plans for Martha, she was so fine and sweet, the apple of his eye— even her young stepmother vowed that Martha was the loveliest character she had ever known. Susan had disappointed him in her marriage, but he would set Martha on top of the world. He could do it, he had bought a large acreage skirting Chouteau Pond on a speculation that would double his fortune; and he was building a mansion on Chouteau Avenue with five tall arches holding up the lofty porch roof. This nonsensical courtship must end.[17]

"Mr. Eads, I will acknowledge to you that I am an ambitious man, and anxious to unite my daughter to some one of the families of the highest standing in the country."

James was dashed. Describing the ordeal in a letter written to Martha later that night, he admitted that his blood boiled for an instant as he tried to break across with the plea that the heads of those very families had less promise of success at his age than he could boast, but he was drowned out:

"And I want everyone to understand who may wish to marry Martha that she will not receive a cent before my death—"

"I consider Martha a fortune in herself," the flushed young man interrupted. "And I am fully competent to support her without the assistance of anyone."

"Yes, yes, all that is very pretty, and is just what I would have said myself in your situation." Dillon went on in a milder tone, conceding that he did not find James altogether hopeless, he was well known for having a long head, he was polite, an admirable son to his parents, and at least not a Yankee like Sue's husband—why, sir, Yankees made a business of coming west to look about for wealthy wives! Well, well, he would give the matter his full consideration and answer it shortly by mail. And he would be delighted to have his caller remain to dinner.

James had drawn himself up in youthful dignity. "No, sir, I am much obliged to you, I must return. Present my compliments to Mrs. Dillon. Good day, sir."

He poured out his hurt and disappointment in the letter to Martha, declaring that a verdict had already been rendered against him. "May God in His infinite mercy guide you from all harm," he ended on a poignant key, "and bring this matter speedily to that happy crisis which will consign you to the arms of him whose greatest pride and most exquisite pleasure will be to 'love, honor and cherish you.' "[18]

A week later the promised answer from Colonel Dillon reached him. The writer, now that the resolute suitor was not at hand, expressed himself to the unflattering full. It was to be like that throughout James Eads's life—not until his back was turned did men say the adverse things they had held in when his searching eyes were upon them. Few experiences had ever mortified him so much, James wrote Martha of this reply. Her father had accused him of "abusing hospitality, of trying to draw off, insidiously, the affections of his child," and had forbidden him ever to step foot in his house again. James implored Martha to marry him at once and let time appease her father.[19]

At a stolen meeting with Martha he pleaded so fervidly that she recklessly promised to marry him with or without parental sanction, thus leaping into a very abyss of unconventionality. But she held to a far-off November date. She hoped that meanwhile she might be able to win over her father, she could not bear to hurt him. Treading on air, James went to see his partners. He would have to withdraw from the salvaging company, he revealed, and cast about for a business on land.

There was doubtless a twinge when he sold them his interest in the Submarine, but he was already teeming with plans for the future. Opportunities were not lacking for a young man with vision and a substantial nest egg, the country had recovered from the economic depression of the late thirties and entered upon an era of restless development. It reached into wide undertakings, longed for more room, for farther frontiers. Appalled as it had been by its own vastness after the purchase of Louisiana, it now felt its wings beating against its borders. Even the certain annexation of Texas within a few months did not satisfy the national hunger for territory. Politicians, playing upon the general hankering for California and its long Pacific coast, harped upon the "manifest destiny" of the United States to stretch from sea to sea.

Statesmen claimed the Oregon country as far north as Russian America (Alaska). The fitful surges of emigrants westward had steadied to a stream, farms sprang up on the prairie overnight, towns grew about once lonely boat landings.

One of the chief needs of this thriving middle country was glass, James Eads concluded, and none was made west of Ohio. And excellent sand for glass had been found in Missouri. He put the proposition of a glass factory up to his late associates, Calvin Case and William Nelson. Glass, glass—when Jim Eads talked of it, surely the most important commodity in the land was glass. He signed the two into partnership with him and set off to the East to look into the processes of glassmaking.[20]

His spirits were high. He wrote Martha jauntily on the way: "We started from Pittsburgh . . . who are we? Why, me and my trunk and umbrella." He covered pages describing the sights, all new and wonderful to him. The one that impressed him most was a church overflowing with worshipers, numbers kneeling in the churchyard, a devotion that awed and exhilarated him. His letters were ardent, he begged Martha to speak more openly of her love for him, he deplored the time already "lost forever" that they might have had together. He was "so wearied" of being away from her and so impatient to get back to her that nothing moved fast enough for him. "Oh, how I dislike this late rising!" he complained. In Quebec he was especially irked with the leisurely air and late meals—if this was a sample of Old England, he said, he would not be surprised "if they don't dine there until the next day."[21]

In September he hurried home, met Martha at the home of her sister, Sue Stevens, told her of his good prospects and begged her to marry him without any more useless delay. She consented, fixed a date a month away, and they exchanged daguerreotype miniatures. There was much furtive planning, Martha's little gales of qualms, James's sturdy insistence, the aroused suspicions of Colonel Dillon, his young wife's sympathetic aid to her stepdaughter. Whether Martha's father became reconciled to the marriage at last is not recorded, but on October 21, 1845, there was something of a wedding at the small Cathedral Church of St. Louis de France, at Second and Walnut streets, almost overhanging the river.[22] It was a short walk down to the waterfront where, with bells ringing and whistles screeching, the young couple boarded a steamboat, waved farewell to the gay party ashore, and were on their way up the Mississippi to Argyle Cottage, where they would make their home for a while.

CHAPTER III

THE BELL-BOAT FLEET

AFTER a too brief stay at Argyle Cottage, James Eads left his bride with his parents and returned to St. Louis, where he began to establish his glassworks in a leased building at 2300 North Broadway.¹ Everything went more slowly than he had expected. All equipment and skilled workmen had to be brought from Pittsburgh or farther east, and when his orders were delayed or ignored he journeyed up the Ohio and took a hand in filling them. To a salvage diver a steamboat ride did not loom as a perilous adventure, but he assured Martha that he was taking all due precaution. She worried about his being on the rivers so much, she had thought he was through with that. "Maria's bustle lies in the upper berth all blown up, ready for use at the shortest moment," he wrote her from the steamboat *Mail*. Whenever he could settle long enough in St. Louis he had Martha come down from the little pioneer town, which had changed its name from Parkhurst to Berlin, to stay with him. His plain boardinghouse room, a block from the glassworks,² was a corner of heaven while she was there.

They had endless things to talk about, the factory, the river, the shuttling scene of life around them. James took a real pride in the progressiveness of his city, though he could find in it some angles for mirth. The waterfront milled with activity as far as eye could see, its babel punctuated by the new boat whistles that had replaced the earlier gun signals—their shrillness jumped the passengers from their seats on deck and froze the wharves for a startled moment.³ There was an excellent municipal water system, as why should there not be? The Mississippi carried past here a million and a quarter gallons of water per second, and a mammoth Indian mound reared above the town to support the reservoir. The water, piped into the very homes, was acclaimed as unexcelled—one had only to set his glass of the rich brown fluid down, wait for the mud to settle, then sip warily off the top of it. "Persons soon became very fond of it," a local historian wrote, "preferring it to any other. . . . The sweet pleasant taste renders almost any well or spring water insipid by comparison."[4] And there was a horse-omnibus line that had been established by James's young friend Erastus Wells, who had come from the East two years earlier with his savings of more than a hundred dollars, secured the aid of Calvin Case, one of James's "wrecking" partners, contrived an omnibus by setting a box top on a spring wagon, and driven it himself. Now he had several buses jogging importantly in and out of the plodding wagon streams.[5]

The omnibuses furnished a convenient way of getting around, and James and Martha had many places to go. There were visits to Susan and her Yankee husband, Dr. Stevens, to Colonel Dillon's, to their friends. There were dances, the statelier of them held at the Tobacco Warehouse and opening with a venison supper. There was the theater, no longer confined to the old Salt Warehouse as in earlier years. James H. Hackett had lately appeared in St. Louis; also two rival Shakespearean actors, the vigorous American Edwin Forrest and the subtle British William Macready whose hissing of each other's performances had caused lively theater disorders on two continents.[6]

James escorted his pretty, spirited wife about proudly, but he found less and less time for it. Trouble haunted his business, his partners had withdrawn from it and he had to cast around near and far for loans to carry it on. Money was growing tight, for the country had been caught up in a crisis with Mexico. The overland trail from St. Louis across the southwestern plains to Santa Fe, a sprawling town of adobe houses, dirty streets, gay, careless people, sleepy little mules and pariah dogs, passed over ground that was disputed between Texas and Mexico. Traders using it were in constant peril. It was an important road, it carried a wealth of commerce, it fed the steamboats, brought gold and silver, and must be protected at any cost. The whole country was already in a belligerent mood, a grievance was not unwelcome. War . . . land, more land, new shores, throbbed the rising undercurrent of expansion.[7]

The young industrialist worked earlier and later, begged credit from St. Louis to New York, combed the eastern markets for factory pots, fire clay and pearl ash, fenced with labor troubles. He found that he would have to give the art of dealing with men as careful study as he had the art of glassmaking, and worked out a policy of justice, good nature and firmness. But in the end he would be very proud, he confided to Martha, to have set up the first glass factory in the West. His only deep complaint was of being separated from her, and that their visits back and forth were so snatched and hasty. He hoped soon to have a home for her in St. Louis, he feared that she was overworking in her solitude for his parents.[8] Then, all at once, the brick-paved streets of St. Louis rang to the tramp of marching men. Spring of 1846 had thawed the rivers, swept the snow from the plains and ushered in war.

Supplies from the East were harder to get, but by the end of summer an encouraging amount of "flint glass" was emerging from the factory, local orders were filled and a shipment made to New Orleans. The newspapers made much of this. Young Captain Eads had become a personage, indeed. He was warmed by the commendations and jaunted off in high spirits to Argyle Cottage, carrying some of his choicest bowls and glasses. The world was almost in his grasp. And presently a bit of heaven had fallen into his lap—in August, 1846, Martha gave birth to a daughter. James planned enthusiastically for Martha and the baby, little Eliza Ann. Every comfort should be theirs—he had only to bring up the glass output of his plant, an eager market awaited it.[9]

In the year that followed, it seemed to James Eads that he had never been so wildly happy or so frustrated. There were weeks when Martha and the baby shared his room in St. Louis, there were his hasty visits to Argyle Cottage and all that it held. But successive petty crises afflicted his glassmaking, pots broke, whole batches of glass were ruined, his men quit. His own and his father's debts hounded him. Still he found time to do some shopping for Martha and his mother, to hunt a boy to help his father, and to gather an occasional anecdote. "Saw your father's man Patrick 'the day,'" he wrote to Martha. "He got married a few weeks ago on Monday morning and on the following Saturday was presented with an heir ... he is highly pleased."

Slowly his plant swung into better production; then, just as success dangled almost to his grasp, the uneasy times squeezed his market dry, the unwanted glass glutted his storage space. Desperately he tried to stave off failure, pushing it back month by month. But in time of war people could forgo the luxury of table glass, steamboats and caravans were loaded with ordnance, troops and food. While essential businesses prospered all around him, James Eads closed the doors of his plant. It was a hard blow to him, his high hopes were crushed, his pride hurt. He was twenty-seven years old, his savings were wiped out, and a debt of twenty-five thousand dollars weighed upon him. There was but one thing for him to do, go back to salvaging. He borrowed fifteen hundred dollars from his creditors, bought back a share in the Submarine, which now belonged solely to William Nelson, and, early in 1848, was out on the brown waters again, diving for sunken cargoes.[10]

The petty annoyances that had bedeviled him on land had no place here, the rhythm of the water smoothed them away. He belonged to the rivers. But he was away from Martha more than ever, and his letters were cries of loneliness. He begged her to write to him as affectionately as she could. He longed for the baby, he was anxious about his mother, he wanted his father to feel secure and free of worry. Colonel Eads, despite the trouble that his unhappy business affairs caused his family, must have been a very lovable man. James was devoted to him, and Martha mentioned him with more fondness than she did the other and more cantankerous colonel, her own father.

The Submarine found more work than it could do. The war had strewn the river bottoms with wrecks, and the gold rush to California was adding to them with tragic lavishness as men dropped their plow handles, trowels or ledgers and made for St. Louis to begin the long journey up the Missouri River and across the plains. Steamboats were crowded, decrepit tubs were sneaked back into service, accidents mounted. More bell boats were needed, and better ones. James Eads had handed sketches for them to his partner, William Nelson, and now Submarines Number Two and Number Three were being built.[11]

Meanwhile he kept old Number One busy every daylight hour. He had to, his debts lapped at him like a rising tide. Things went well and things went ill. In March of 1848 he was working in twenty-five feet of water when some of a wreck's cargo, casks of bacon, escaped through a broken deck and floated off. It was raining and the wind blowing "as cold as charity," but the yawls had to be manned, the casks chased and rolled ashore, men wading over their boot tops. "As it requires little short of a hurricane to keep me from working," James wrote Martha, "You must know we were stirring the freight today."

The jobs took him farther and farther southward, but he hoped to get back to Argyle Cottage by August, in time for Martha's second confinement. His letters breathed of love, of hunger for affection, of an exalted religious faith, of a hope that they would both live up to their highest ideals. And usually a bit of nonsense was tucked in. "I have been on the eve several times of applying to a mesmeric clairvoyant to ascertain whether Sue will be an uncle or an aunt," he wrote. As for himself, he hoped he would be a father and not a mother, he did so want a boy. He tossed in this riddle: "Why is a cigar like a lady's bustle? Because it is ter-back-her."

He was working near the upper tip of Louisiana, and hailing each passing boat that might possibly drop a letter off for him. No word from Martha had reached him for many days. He had found that he could not get home for her ordeal, but he sent her a gift of a length of dress goods. The critical time passed, and still no news. Anxiety tore at him, the need of money bound him, he felt that he was losing "the very flower of his happiness" by the long absences from home. As his alarm mounted he wrote rebelliously about the hardships of childbirth, blaming most of them upon the tight-waisted fashions that had remolded the forms of women.

The silence grew unbearable. There was a slight chance that his mail had been held up at Cairo. He left the Submarine and went there as fast as a steamboat would take him. Sure enough, Martha's letters were there, a sheaf of them. He read them devouringly, the latest ones first. A son had been born! On August 11, 1848. And named for him.[12] Martha was safe! She had been able to write a poem about "the babe like a blossom unfolding to light."[13] Tingling with relief and joy, James covered page after page with small close writing, then turned the sheets around and wrote across them the other way. But he had to set back southward to the Submarine, and complained, "If any poor fellow ever wished to get home it is me."

Near the middle of October, after four months of absence, James wrote that he could get home for a while, and broke into rollicking verse over it. "Twas to win and hold the captain bold" that she was given her charm, he teased Martha. He was coming home, *home,* and she must "put her sewing out and get a nurse for 'the childer.'"

Argyle Cottage sat triumphant in a blaze of crimson and gold autumn foliage as if aware that it sheltered priceless treasure, a young man's wife, his tiny daughter and infant son. It was a reunion never forgotten. He had a son, the image of him, and with his very name. It was hard to let Martha and the babies out of his sight, but he tramped patiently about the town with his father, who was full of ambitious plans. Berlin, the once Parkhurst, had joined the nearby village of Le Claire and adopted its name.[14] The possibilities of expansion were remarkable, steamboatmen and Rock Island rapids pilots were moving here, new town additions had to be laid out, houses built. All that was needed was money. James arranged credit, paid bills, and all too soon his stay was over and he was out on the water.

Shortly the new Submarine Number Two was completed and proved almost excellent. It could lift a modest-sized wreck right out of the water, saving the boat as well as the cargo. Now James Eads's salvage duties took him from one of his bell boats to the other, he toiled in deep water and shallow, under hot southern sun or gray northern skies. In February, 1849, he was working in the Illinois River, the air and water frigid. He had much trouble placing chains under a wreck so that it might be lifted and restored, and had just succeeded when a run of ice bore down on it, turning the submerged hull over in the chains, leaving it useless. The misfortune touched off his homesickness, his letters throbbed with it. "Kiss the boy a thousand times for me," he begged Martha.

Spring opened slowly. The last of the ice had floated away and vessels in new coats of paint were teeming the river when an unprecedented disaster befell steamboats at the St. Louis wharves. About nine o'clock on the night of May 17, 1849, the steamer *White Cloud* caught fire, burned its hemp mooring cable, drifted around in a capricious current, collided with other boats tied up at the waterfront, igniting them. Set loose, they rocked about, blazing, toppling, sinking. Sparks blown inland set warehouses on fire. The volunteer firemen turned from the doomed boats to save the town, resorting to gunpowder blasts to check the roaring flames. Twenty-nine steamboats were destroyed in that holocaust.[15]

Steamboatmen were sickened by the loss. They had pride and affection, as well as money, invested in their boats. Whatever depression James Eads may have felt from this calamity was wiped out a month later by a personal tragedy. His infant son, James, never strong, flickered out of life on June 15, 1849.[16] The woods around Argyle Cottage were joyous with the song of birds, but silence lay as a pall over the friendly house. Martha appears to have borne her grief with a strange sweet valor. James was stricken to his soul over his loss and regret that he had seen too little of the child he had adored. It was to be a long time before the heartbreak over his son's death ceased to haunt his letters. He could not stay now to comfort Martha, his earnings were too much needed—he had promised to help his father over some financial hurdles during the next winter. The rivers reached out and claimed him.

They demanded more and more of him, he traveled the streams almost constantly, exploring each drowned hull, planning each job, performing the critical work in "dangerous and exposed places where his men were unwilling to go."[17] His visits home were snatched tastes of bliss. He poured out his miss of Martha in long letters after leaving her, wooing her as ardently as in the early days of courtship, selling himself frankly as a right good sort of man—there might be better husbands, he wrote her, but the chances were three to one against it. As the winter of 1849-50 set in, he finished salvaging a boat in the Ohio, and was soon pinned to a difficult task that he had undertaken for the city of St. Louis, the removal from the harbor of some of the boats that had sunk there in the fire of last May. It would tax his Submarines to their utmost. In fact, it called for a bell boat such as had never existed, and he was designing one in whatever time he could spare for it. He would not be able to get to Argyle Cottage for Christmas. " 'Drive on' is my motto," he wrote.[18]

It was a year later, 1851, when Submarine Number Four materialized. Provided with powerful Gwynne centrifugal pumps to rid a hull of water and sand, and with the stoutest of hoisting equipment, it could raise the largest steamboat and set it afloat.[19] The feat was acclaimed as sensational, and Captain Eads, then thirty-one years old, was hailed as the miracle man of the waterways. He operated the new Submarine himself, from St. Louis to the Gulf, walking the stream's uneasy bed, directing the pumps, securing the chains. It prickled him with satisfaction to see a submerged monster heave up through the surprised torrent, dripping sandy water from its forlorn decks. Tugs were ready to pull the rescued craft away, and before many weeks had gone by, the skipper of Number Four might meet the resurrected boat, painted, refurbished, sailing the streams as proudly as before.

Another wreck and another, yet the diver-captain had energy left over for a swim in the evenings "with Patrick and a life-preserver," and long sessions of chess when he could find anyone who did not mind being beaten repeatedly. He crowed over the discovery of a skillful player, a passenger he met on a steamboat, who "bore defeat better than any chess player I ever met." He found time to send home gifts, a Bible and lounge for his mother, table covers and "handsome window shades" for Martha, gooseberry bushes that his father wanted, and always money for the household budget or to feed the maw of his father's debts. He took time to go to Colonel Dillon, who was dying—the old gentleman seems to have grown fond of his obscure son-in-law in spite of himself. He snatched moments to do petty errands: Sue had left her mousseline de laine skirt and her earrings at Argyle Cottage and asked him to have them sent to her; she was feeling "dauncy" and he was trying to get her to take a cathartic, but without success. He spent hours with a cousin, a dear, if unsophisticated, young woman, whom he had join him at meals at a Cincinnati hotel. "She is not yet able to eat altogether with her fork," he wrote Martha, "but is getting more genteel every day."[20]

Whenever he could he had Martha come to St. Louis to be with him, arranging it, when possible, so that he would find her in his boardinghouse room when he came in off the river, it gave such an illusion of home. He plied her with questions about life at Argyle Cottage, he told her of the wonderful feats that his bell boats were performing, and read the works of Robert Burns aloud, making much of the dialect. He had her read her latest poems to him—some of them had been printed in the Davenport *Gazette*, whose editor later vowed that "as a poetess she was inferior to no female writer whose effusions graced the magazines of the day." There was a very special one, "Lines to my Husband," inspired by James's gift to her of an Aeolian harp, that called for many readings.[21]

They did a good deal of jaunting about, James eager to show Martha the newest sights. A museum had been rigged up in the old Wyman Hall, catering to every taste—there were Indian relics, Egyptian mummies, the paintings of John James Audubon and George Caleb Bingham, a midget known as General Greene, and the only Female Sax Horn Band in the world. In the music room Jenny Lind gave several concerts, her manager, P. T. Barnum, interspersing them with temperance lectures. Usually a minstrel show kept the stage lighted and the room packed,[22] and no one liked nonsense and a good laugh better than James Eads.

But the outdoor city still offered the richest drama. The waterfront, much of it graded up to the cliff top and paved, and coming to be called the levee, was a colorful melee of movement and sound. The colonnade of boat chimneys had grown taller and longer. Multiple lines of wagons twinkled past it on Main Street or jammed and tangled while drivers filled the air with shouts. Horsemen rode in from the country, cantering smartly past the vehicles. Droves of cattle often overran the sidewalks, scurrying pedestrians into doorways. French and Indian half-breeds, down from the north on lumber rafts, padded about in their lithe way.

There were wistful talks in the quiet of the boardinghouse room about the time when all debts would be paid and there need not be so much separation. There were hours of playing with small Eliza, visits to Martha's stepmother and to Sue's family. Time flitted, one could not stay it. And all at once, his heart sore, his arms empty, James was being borne off to an urgent job and Martha back to Argyle Cottage, the river stretching longer between them, the bluffs thrusting out a rugged shoulder to shut them apart.

James was never downcast long, for life was full of rich promise. Another daughter was born in March, 1851[23]—James named her Martha, but called her Little Mattie. Her coming helped to heal the loss of his son, but she added a new poignant longing in the weeks that stretched between his visits to Argyle Cottage.

Yet there were days so full of peril and strain that his loneliness was held suspended. One such day was to stand out sharply in his memory. During the heavy flood of this year, 1851, he found it necessary to descend to the river bed in a current so swift that he had to devise extraordinary means to sink a bell in it. Uprooted trees twisted and tumbled past, and at the bottom "the sand was drifting like a dense snow storm," he related this breathless adventure. "I found the river for at least three feet in depth a moving mass, and so unstable that, in endeavoring to find a footing in it beneath my bell, my feet penetrated through it until I could feel, although standing erect, the sand pushing past my hands." The next day, driftwood carried his bell boat downstream and there was an ado to get it untangled from the mass.[24]

Important contracts were seeking out the new Submarine Number Four. The salvage income had at last turned upon the besieging debts, routing them. Everything looked bright, James Eads's dreams soared. Then, a few months later, began a series of happenings that rocked his world. In June of 1852, just as urgent work called him from operations at Island No. 82 (above Greenville, Mississippi), to New Orleans, his mother, who had been ailing, grew worse. She lingered a month and died. Martha, whose hands had been too full, drooped with fatigue and grief. As weeks passed and she seemed slipping from him, too frail to grasp at life, James was alarmed. Dr. Charles Stevens, Sue's husband, suggested a rest at the Water Cure in Brattleboro, Vermont. It was September when James arranged to leave his work, sent six-year-old Eliza to Sue in St. Louis, bundled Martha, the baby and Bridget, the nurse, into a carriage at Le Claire (earlier Berlin) and drove to Davenport. From there by railroad and steamboat they reached the Vermont resort.[25]

The fear that had chilled him eased away. Martha had been hardly able to sit up in the carriage when they left Le Claire, now she could take long walks with him through the beautiful frost-tinged countryside. The future had never looked so golden. Martha bubbled with little plans, for they were going to move to St. Louis and have a place of their own and she was full of it.

James had a big plan that had been shaping in his mind for a long time, and someday, perhaps not too many years off, he would begin to put it through—the clearing of the Mississippi channel of every obstruction down its whole navigable length, hauling up old hulls not worth a penny, jerking out every snag, pulling down every pile of drift logs.

Abruptly their stay at Brattleboro was cut short, a childhood epidemic had invaded the place and they fled to Baltimore with their baby. There they picked up James's cousin, Fannie Buchanan, took her along to visit other relatives on the way home, and at Cincinnati tried to get passage on a steamboat for the last lap to St. Louis. The boats were so crowded that they let four go by, and then, as though manipulated by fate, they boarded the ill-starred *Dr. Franklin II,* which had come along fairly empty. An old river acquaintance, Captain Shallcross, put his frail wife on the boat and asked Martha to look after her if she became ill. At Louisville passengers poured aboard, they sat on their bags, slept on the open decks. A cold drizzle set in, a dark fog closed over everything. The river was low, the shoals threatening, it seemed safer to tie up in the Portland Canal that detoured about the Falls of the Ohio near Louisville, and wait for daylight before steaming on.

That night cholera in thin disguise stalked forth on the vessel. Mrs. Shallcross was taken sick, and by morning sent for Martha. It was not until the stricken woman was in her death agony that James realized what ailed her. He took Martha from the room and up on deck. She was distressed by the piteous scene, and he tried to cheer her by talking about the home they were going to make —he would never be away from her so much again, he could afford to loaf a bit, leaving the bell boats to his men.

The steamer grounded on a bar above Evansville, Indiana. It was annoying, no time was too soon to get off this boat. In the late afternoon Martha was seized by the swift plague. There was no doctor aboard. James was frantic. He got some of the boat officers to row ashore and bring the nearest physician, three miles away. All night they fought the greedy, clutching plague, losing to it steadily. By daylight—it was October 12, 1852—Martha, exhausted, was sinking fast, she could not answer when James told her good-bye for the last time. His leaden feet followed as strangers carried her off at Evansville. The baby clung to him, crying feverishly. He rushed her to a doctor, he could not lose her too. She grew better, and the sad journey was resumed on another boat. At Chester, Illinois, on the way up the Mississippi, he hunted up another doctor—it seemed to him that the child was fading from life, too listless to sob longer for her mother. She rallied, and James carried her aboard the boat again.[26]

He brought Martha ashore at St. Louis, her songs silenced. The plague was there ahead of them, physicians toiling over the sick until they fell with fatigue. Wagons had become hearses, funeral hymns were the only music heard in the shocked streets.[27] James followed one of the wagons to the cemetery, turned away from the fresh mound, bleak and empty, paused a helpless while at Sue's where his children were to be given a home, and went back to the river.[28]

CHAPTER IV

COUNTRY GENTLEMAN

THE river, its winter grayness, its summer glint, the silver night sheen as it wound past silent prairies or between dark twittering forests—the waters swayed and rocked, idled off in wide curving sweeps to double back nearly upon the course they had abandoned. They turned out patiently for a shoal they had built in vehement haste and attacked with relentless fury a high, sturdy bank, hurling giant oak and sycamore into the lashing current. The river—it sang in the heart, surged in the blood, it lapped through the crooning of slaves who toiled in the fields beside it, lilted in the swinging chant of boatmen. It stole into the throbbing rebellion of a young salvage-fleet captain whose happiness had been wrenched from him before he had time to taste it fully. It lulled and lured, awed and challenged him, filled long hours with conjecture. Whenever James Eads emerged from his silence he talked of the river, of its beauty and power, its inestimable service to men.

Nowhere else on earth, he often said, was there such an important trunk stream as the Mississippi, or such a vast navigable river system as provided by it and its branches. These waterways were almost the only roads in the great flat valley that reached from the Alleghenies to the Rockies and from the northlands to the Gulf of Mexico, comprising a million and a quarter square miles of land rich in fertility and ores, blessed with a temperate climate. Yet these phenomenal natural roads were taken lightly for granted, exploited, neglected. Little had been done for them since Henry Shreve ceased a dozen years ago to belabor Congress for funds to keep his snag boats running. They needed attention, their channels cleared of driftwood and of even the most worthless wrecks. Surely, men of large shipping interests or of political power could get this done. No one dissented from this, but no one did anything about it.

His work had steadily expanded until it called him back and forth over more than three thousand miles of stream, but he got to St. Louis whenever he could, sat by the hour in Sue's hospitable home, holding his little girls, Eliza and Mattie, or watching them at play. And often he went to see Eunice Eads, widow of his cousin, Elijah Eads. Eunice, fair, self-possessed, her pale chestnut hair rolled back from a classically pretty face, had had a sorry time since her husband died four years or so before, leaving her with three small daughters and little else. James had written Martha in November, 1849, that he had found Eunice "very uncomfortably situated" with her aunt, a Mrs. Williams, who had offered her and her children a home for the winter but had soon shown an absurd jealousy of her, and had rented the house to a woman with whom they all boarded. Eunice had scant funds for paying board, and she dared not accept help offered by her aunt's well-meaning but puzzled husband for fear of precipitating a scandal. James had advised her to seek another boarding place at once and he would see that she could afford it.[1] Eventually his cousin Eliza, Colonel Dillon's widow, had taken the forlorn family in, with James paying part of their expenses. Later he confided to Martha that he believed Eunice was about to marry again, but nothing had come of it and she still eked out a cramped existence with her little brood. Now her plight and her dependence upon him for guidance were creeping into his lonely heart. Her children, sprightly Genevieve and Adelaide and quieter Josephine, all blond and chubby, two of them a little older and one younger than his own seven-year-old Eliza,[2] looked to him as a father.

A year and a half after Martha's death, James Eads's health gave way. Years of hard work, exposure, loneliness and grief had at last battered down his resilient endurance. His doctor warned him that he would have to put aside all work, only complete rest could save him. In need of reassurance, he turned to Eunice. She was practical and comforting, she was calm and beautiful—when he was with her it was as though a storm-racked boat had found shelter in quiet waters. On May 2, 1854, James Eads and Eunice Hagerman Eads were married at St. Vincent's Catholic Church, on Ninth Street, not far from the Dillon home.[3] Cousin Eliza Dillon, who had adored her stepdaughter, Martha, was with them, signing the marriage certificate as witness. Leaving the older children at the Convent of the Sacred Heart on South Broadway —for Eliza Dillon's hands were full with her own children— James Eads journeyed east with his bride to sail on the first of his many trips to Europe.[4]

The ocean voyage was invigorating, although afflicting—James Eads was a poor ocean sailor then and ever after. The panorama of foreign lands relaxed him. He delved with his musing curiosity into all that he saw, storing many an odd treasure away in his mind. There were rivers picturesque and neat, none of them as vast and plunging as the Mississippi, there were craft of many types, none as majestic as the broad valley steamboats with their mounting decks. There were historic cities and quaint villages, colorful peoples, absorbing art treasures, restful countrysides, and always Eunice, sympathetic, watchful for his comfort. His health returned slowly, and in a while he was ready to go home. He felt quite fit again, he insisted, and wanted to get back to salvaging.

Returned to St. Louis, his first concern was to get a home and gather his family into it. He bought a stately brick house set in a stone-walled yard, on Fifth and Myrtle streets[5] and furnished it. It was a pleasant, comfortable home, and Eunice, of Dutch ancestry, was a good housekeeper. But James did not have much time to spend in it. He was back on the rivers and they demanded him more than ever.

Besides hoisting cargo and sound boat hulls, he went to much expensive trouble to take decrepit craft out of the channels. He was beset by the urge to clean out the streams, he could not shake off their mute appeals for the help that no one seemed minded to give them. President Pierce, who had come into office a year before, cared less for the nation's water roads than had any executive before him, his consuming interest appearing to be a scheme to buy Cuba and perpetuate slavery there.

Indeed, so unconcerned was Washington officialdom with the inland waterways that, in 1855, it laxly offered five of the government snag boats for sale. James Eads was amazed—the snag boats set adrift, perhaps broken up, and the rivers left to their fate! Full to the brim about it, he went to his partner, William Nelson, builder of the Submarines, and proposed that they buy the snag boats. They must be kept running, the condition of the navigation channels was disgraceful and growing worse. How would they finance the profitless undertaking of pulling up drift logs and worthless boats? Eads did not know, but he would find a way.

They bought the snag boats for a hundred and eighty-five thousand dollars and began to convert them to bell-boat Submarines, which could draw up any sort of obstacle.[6] Still the question hovered: How could the cost of channel cleaning be paid for? James Eads spent hours over a chessboard as he mulled this. He played with terrific concentration, partly, perhaps, as an outlet for the emotional energy that mounted in him when he was gripped by a strong intention. He liked to make the game more difficult for himself by playing without seeing the board, calling the moves, carrying the positions in mind. Or he would play several games at once, each with a different opponent[7]—it was like that in life, one could not hope always to fence with a single adversary, he had to meet all comers. He played relentlessly, bent upon keeping his opponents from taking a single pawn. One day, early in 1856, he started for Washington with a bill he had drawn up to present to Congress, offering to clear the neglected valley streams and keep them clear for five years at a reasonable fixed annual sum.[8]

He found the Capital a caldron of dissension, anti- and proslavery factions trying to shout each other down. There were cries of states' rights . . . free soil . . . squatter sovereignty . . .

the dying gasps of the Whig party, the infant wails of the new Republican party. Would no one listen about the valley rivers? The streams were choking, struggling sickly along in their littered beds in low-water seasons, raging over their banks at every flood to lay waste the countryside. Crews and passengers were going down to death, sacrificed to inertia, to laissez faire, valley wealth was sinking in the greedy waters.

Of all the lobbyists in the Capital, the slender, ruddy riverman was the most indefatigable. Up and ready for work before the city had turned over for its last nap, he was everywhere present, courteously but stubbornly insistent. The Mississippi, the nation's greatest road, was strangling with debris, he pointed out. The Ohio, Missouri, Arkansas and Red rivers, vital trade routes, filling with logs, shoaling around wrecked craft—was western commerce to die? Legislators turned distractedly from their heated feuds to listen to the persistent Westerner, they could not well escape him. The rivers, the rivers, he never let up about them. The bill he fostered passed in the House of Representatives but was defeated in the Senate by the opposition of Jefferson Davis, who believed that it would be a mistake to take up the proposal of a person "whose previous pursuits gave no assurance of ability to solve a problem in civil engineering . . . which involved the control of the mighty rivers."[9]

This was James Eads's first major defeat and hard for him to swallow, but the game was not lost because a move had been blocked. On his way home he worked out a scheme for carrying on the river clearance, and had hardly more than stepped ashore from a steamboat at St. Louis before he was organizing a Western River Improvement Company. His visions were always limned on a broad canvas—he would have this company chartered to operate anywhere in the world.[10]

He took his outlined plan to marine insurance companies all over the valley. The time to save a boat was before it went down, the salvager argued, as if trying to put himself out of business. Clean channels would save many an appalling loss-payment. Reduced losses would lower insurance rates, he reminded steamboatmen. Low insurance rates would cut freight costs, he told shippers. Clearing the rivers would not be as heavy a task as it appeared, for the channels seemed more cluttered than they actually were because their currents heaped brush and silt about every obstacle, building it up and out until the waters had to swing around it. He would illustrate this with some diverting tale of his own experiences, often with the story of the island that had reared over the steamboat *America*. He had always been able to talk men into doing things if they would listen, and he now signed more than fifty insurance firms into his company.[11] Vessels and cargoes could be made fairly safe on the rivers, he assured them and boat-owners confidently.

As if waggishly to refute this enthusiastic claim, the Mississippi lost no time in staging the greatest of all steamboat tragedies, pointing it with especial irony by starting out with Submarine Number Four as an instrument of destruction. James Eads had just finished using the bell boat to raise the steamer *Parthenia*, sunk north of St. Louis, and had moored it above the city. It was midwinter of 1856-57, severe cold soon froze the stream from bank to bank. As usual, the west shore was flanked with craft. "The riverfront," reported a watchman on one of the doomed vessels, "was so solidly lined with steamboats that one could walk upon their decks a distance of twenty blocks."[12]

An unseasonable thaw far above St. Louis broke up the ice there and sent the weaving massive cakes downstream. When they finally drove against the unthawed surface near the city, they pushed over and under it, erupting in piles twenty to thirty feet high. An ice gorge! Alarm zigzagged through the streets, men made for the wharves. The grinding mass had moved down against the Submarine, flailed at its deck and tall derricks, turned it over and pushed the groaning hull down against the *Federal Arch*. The two boats shoved along, tore loose and carried ahead of them every steamboat in their path, the ice surging over them. Bells rang frantic appeals above the roar of the gorging ice, the deafening reports of snapping timbers and the bellow of crushed hulls. The writhing tangle staved on until the Lower Dike halted it. Forty steamboats perished in the disaster.[13]

Many a riverman's heart ached as his vessel, crushed under the still seething ice, moaned as if in mortal pain. They needed help, where was Captain Eads? When Eads arrived he stared gravely at the indescribable ruin. His strong bell boat Number Four lay in the mass of wreckage, but its sister Submarines would be brought. Only a boat here and there could be saved, and that by slow, perilous effort.

The first of the boats Eads recovered was the *Garden City,* after ten days of nearly incredible labor. The steamer, lying somewhat apart at the upper wharves, had sunk until its deck was just underwater, the keel resting on ice packed sixteen feet deep. Ways were laid down under the hull and the boat dragged ashore.[14] The work in the harbor went tediously on, but James Eads did not find it irksome, for he was learning at first hand about ice. Ice was more than a navigation nuisance, more than a ruthless destroyer, it was a designer, creator. The gorges of a receding ice age had helped to cut the river's groove in the limestone bed of the valley, it had helped shape the size and trend of this magnificent stream. The ice gorges of today swept the stone bed clear of sand, in places, and scraped the groove a fraction deeper. Ice and the river bed—he might have need of this understanding sometime.

By spring of 1857, most of the snag boats bought from the government had been put in condition. One of them, Submarine Number Seven, formerly the snag boat *Benton,* was a matter of much technical pride to James Eads, who had fitted it with two strong centrifugal pumps and his most improved hoisting apparatus. He sailed it down near Natchez and brought up the sunken steamboat *Switzerland,* with its cargo, while plantation darkies watched furtively from the high shore and later wove the strange incident into their moaned chants. There were ten Submarines in the fleet now, and Numbers Eleven and Twelve would soon be building.[15] Added to their pumping and hoisting facilities, Eads had designed and patented a device to blast rock underwater, so that even the most obstinate ledge or boulder could be eliminated.[16] He watched over the whole fleet, explored the most dangerous wreck locations, hurrying from one scene to another, not pausing to consider how tired he was until he could go no longer. He was like a man storm-flailed and limp when he suddenly gave way, his stubborn will unable to push him farther. His doctors were stern with him this time. He had never known how to work any way but hard, from dawn to dark, now he would have to retire to absolute idleness if he wanted to live.[17]

Idleness! Desert the rivers! He might have to keep off them, but he would never cease to plan and fight for them, particularly the Mississippi. Thirty-seven years old, and his working days over! Fortunately, his family would want for nothing—he and Nelson had accumulated a half million dollars in the past few years with their fleet of Submarines. There was one rich cargo that this enforced loafing would bring to the surface: he would be with Eunice and their daughters, indulging his long-cherished dream of an idyllic home.

He looked about for a retreat, a place with a garden in which he might putter around, and learned that "a mansion with park-like grounds" out on Compton Hill in the south end of the city, beyond Lafayette Park, was for sale. He went to see it. It was an ideal place—James Thomas, part owner of a private bank, had built the spacious house and landscaped the nearly four acres of ground about it—and the price was modest enough. With no loss of time Eads closed the deal.[18]

If he must idle and loll, twiddle his fingers and be worthless the rest of his existence, this was a pleasant place for it. He laid out a flower garden, built a rose arbor, fenced off a vegetable garden in a large vacant plot adjoining his grounds, installed some cows in the rest of it and gathered his books to a spacious library off the long impressive drawing room. For the most part he was contented. After years of haphazard drifting he had come to anchor in this peaceful haven. He enjoyed seeing Eunice bask in her new setting, not a little proud of being the mistress of the handsomest establishment in the town.[19] His grounds and house rang with merriment as five little girls or young misses, in short-waisted frocks with pantalettes showing beneath, romped and raced.

His home soon became a favorite gathering place for men of many callings. Saddlers or carriage horses beat up the dust or plunged through the mud of unpaved Lafayette Avenue to swing around Compton Avenue and into his driveway that curved under artfully planted trees. He gave all comers a cordial welcome, some excellent wine, often a lavish supper. Eunice, who wore at her waist a chatelaine of keys to storerooms of food, linen and silver, had early learned that, her husband's hospitality being as impulsive as it was, a full meal might have to be served at almost any hour.[20] Over their glasses or plates, rivermen, industrialists, bankers and politicians talked of various things: of a new City Market that had been set up on Broadway, of the annual Fair of the St. Louis Agricultural and Mechanical Association inaugurated more than a year ago, in 1856—the mile-long race track at the Fair Grounds was the finest in America; of the hard times that were growing harder—if Captain Eads and several others had not come forward and given written guarantees to St. Louis people of their money on deposit, the banks of the city would have suffered a ruinous run on them.[21] There were discussions of the recent Dred Scott decision by the United States Supreme Court, establishing that slaves were property and not persons. However distasteful the decision was to James Eads, it had well pleased his cousin, President Buchanan, comforting him for an unexpected turn in the new territories, Kansas and Nebraska. Buchanan had been in favor of squatter sovereignty for them when a bill had come up in Congress three years before to shape them from the unorganized remnant of the Louisiana Purchase, not foreseeing that so many of the wrong kind of people, abolitionists, would settle in them. He was morose about it now.

But mainly the callers at Compton Hill talked about the river upon which nearly all St. Louis interests depended. There were several short railways in Missouri, and the Baltimore & Ohio Railroad, by means of several links, had reached the east bank of the Mississippi over a year ago, but steam trains played a meager part in the commerce here compared to that of steamboats. James Eads's friends were convinced that only he could answer the most vital questions about their waterway, only he dared to encourage their boldest hopes. Over and over they declared that the Western River Improvement Company which he had established had been "an outstanding rebuke" to the government that had so neglected the nation's busiest roads.[22]

Queries, plaints, theories and suggestions pattered around him. Shrewdly he sifted them. With his health so broken, his opinion was about all he had left to give the Mississippi. It was a stream of life, its future commerce would stagger the imagination, he predicted. It was the patron and servitor of a rich land of promise, a prodigal gift of nature. Those along its way owed it a boundless allegiance, they should not fail it.

CHAPTER V

THE FIRST IRONCLAD GUNBOATS

THE autumn of 1860 lingered over Compton Hill as if reluctant to leave. The western sun bronzed the foliage and gilded the tumbleweed that had fled the wind to take shelter in shrubbery and fence corners. James Eads, ruddy with returned health, walked thoughtfully among the rose hedges. He was forty years old now, his hair was receding from his forehead, a pointed beard lengthened his stubborn chin, a shaved straight upper lip was ever on guard against the impulsive lights in his deep eyes. Everything was sharply objective to him in the Indian summer haze, the familiar paths, the trellised vines—the thrush had deserted its nest in the syringa bush. The year was waning, and things might never be like this again. A breathless threat of change hung over the land.

Late afternoon often set up a rhythm of hoofbeats and scrape of wheels on the curving driveway. People were warmly or curiously attracted to James Eads. He was genial, witty, and sometimes startling—at any moment, across his most trivial badinage, he might toss a sentence that would open up a whole realm of conjecture to haunt one afterwards. If his callers trooped to the garden to find him, he likely presented nosegays to the ladies and boutonnieres to the gentlemen with the air of a cavalier, in the elaborate fashion of the times, and led them back to have iced wine on the veranda or in the gracious drawing room. Invariably the talk moved around to his most consuming interest, the river.

The Mississippi, he said, was the grandest physical feature on the continent, and the most useful to men. It drained and enriched an area as large as all Western Europe.[1] Its waters swarmed with craft, southbound boats bearing hay, cattle, grain and factory products saluted northbound boats heavy with sugar, cotton, oranges and turpentine. Passengers, rich and poor, free and slave, filled the staterooms or huddled on the open decks. The New Orleans waterfront was one of the busiest scenes of trade in the world, and the St. Louis levee only a step behind it. The steamboat arrivals here in the past year numbered more than thirty-six hundred![2] Nothing, James Eads held stoutly, should be permitted to disturb the security of the great stream.

At times he drew aside with some of the callers to talk in low tones of the country's unrest. Even the election, in the past weeks, of Abraham Lincoln, a Liberal Republican—that is, an ardent Unionist but not a rabid abolitionist—had not begun to bind factions together. The nation was drifting to a crisis. James Eads, who had never owned a slave, had left the ranks of his proslavery kinsman, President Buchanan, and aligned himself with a sturdy group of St. Louis Liberals. Among them were big, red-haired, honest, profane Francis Preston Blair, who was gradually freeing his slaves, and his wiry, red-haired, forceful cousin, Benjamin Gratz Brown. There was James Rollins, father of the State University, a tall, commanding Virginian with inspired, farseeing gaze and resonant voice. And Edward Bates, a small sized man of towering personality, and so popularly known for his integrity that he would likely have been made president instead of Lincoln if the Republican convention had been held in St. Louis, not Chicago. Often on hand, and in general agreement, was a Democrat, Eads's old friend Erastus Wells, of penetrating mind and pleasant manner, whose bus line had expanded to a thriving horse-car rail system despite public protests against the menace of such speedy vehicles in the streets. Wells, always a stride ahead of his townsmen in civic matters, was busy with a revolutionary plan to reform police regulations so that a ruffian element of "plug-uglies," such as infested every city of the land, could be quelled. But he was deeply concerned with saving the Union, too.[3] Secession was driving its wedge, James Eads kept saying, and when the country was split across, the Mississippi would be cut in two, the lower half of it closed off from the North. The only thing that could save the river was a navy of its own—gunboats, armored gunboats, able to endure the fire of the shore batteries they must pass. Gunboats, Eads insisted, river gunboats.

His intimates agreed with him, but when his idea of river protection seeped out through the town it brought many an amused smile. A river navy! Iron-covered boats, as if there were such things! One of the delightful characteristics of Captain Eads was his soaring imagination. Warships on the Mississippi—what would he think of next? Anyway, all this wrangle over secession was likely no more than an oratorical bubble that would presently dissipate.

It was not until midwinter of 1860-61 had shut the world off from the icebound port that St. Louisans, turned inward upon themselves, filled the air with their exaggerated differences. The very house in which James Eads and Eunice had spent the first three years of their married life, on Myrtle Street, was a point of contention, for its new owner had turned it into a slave pen and market.[4] Eads watched antagonisms heighten, Unionists and Secessionists clash at his fireside, cleavages show in business firms, churches and families. He grew more alarmed when, on December 20, 1860, South Carolina seceded from the Union and seven other states followed it in quick succession. Missouri was rocked. Predominantly southern in sympathy, northern in geography, "a peninsula of slavery in a sea of freedom," which way was it to turn?[5]

On February 4, 1861, a frail hope of smoothing over the country's dissensions was held out by the convening at Washington of a Peace Congress. But the puny pleas for peace were drowned in the country-wide tumult, Missouri furnishing a good part of the noise as it tried to argue itself into one camp or the other. A convention was called at the state capital to make a final decision, the delegates assembling on February 21 "in a small, repulsive Courthouse." The weather was vile, the oil lamps hardly more than outlined the muddy streets. The townsmen seethed with hostility to the Unionists. A delegate from St. Louis, Dr. Linton, reported: "When I got to Jefferson City and heard nothing but the Marseillaise and Dixie instead of the Star-Spangled Banner, I felt uneasy." Others were uneasy, too. The convention was moved to St. Louis and housed in the beautiful Mercantile Library.[6]

The body was overwhelmingly proslavery, but wary of secession. Even Sterling Price, later to be a Confederate general, was firmly opposed to Missouri's quitting the Union, for it was a grave question how a seceded state hemmed in by Union neighbors might fare. The debate went on in bitter but cautious earnest while hotheads of both factions muttered in the streets that no convention could decide for them. But the convention did decide, and against secession. A roar of jubilation and protest shut out all explanations.[7]

Into this din was tossed, on April 12, 1861, a deafening bulletin: At daybreak southern General G. T. Beauregard, of seceded North Carolina, had ordered his shore batteries to fire upon Federal Fort Sumter in the Charleston harbor. A battle was on. It raged all day and night. On the 13th, the fort, its ammunition exhausted, fell to the enemy. The distant shots had seemed to echo along the Mississippi, through ports and across fields. Bawled threats and shouted defiance rose from street corners. Farmers hurried home from crossroad stores to clean their muskets or load shells.

The cherries were in blossom at the Eads place on Compton Hill, the sun filtering through their white lace to dapple the pale new grass. The Union flag lowered! War. It was so near that it squeezed the breath. War, and the cherries abloom—when one looked through their waxen beauty at the blue sky it was unbelievable that men would fall upon each other in murderous conflict. War, and there was not a single Union gun on the Mississippi or any of its tributaries! It had been a barbing fact to James Eads for months that the South had recognized the importance of the rivers and had seen to it that strong forts were built along the half they meant to hold when the inevitable break came, while the North had dallied. Even men who agreed with him that a Mississippi navy was vitally needed had done nothing about it.

He began on them again: gunboats for the Mississippi, armored boats that could endure the fire of shore batteries, ironclad boats, a fleet of them. He wrote to Edward Bates, who had been made attorney general in Lincoln's Cabinet. Vigorous action must be taken, he said, a defensive war was not enough. Bates's reply electrified him: "Be not surprised if you are called here suddenly by telegram," it said cryptically. "If called come instantly. In a certain contingency it will be necessary to have the aid of the most thorough knowledge of the western rivers and the use of steam on them."[8] The gunboats were looming! It would do no harm to pack at once.

At that, the packing had to be finished hurriedly, for the telegram was on the very heels of the letter. In a family flurry James Eads was off to the Capital. The trip was annoyingly slow. Railroad schedules were already disrupted, lines threatened, Baltimore was cut off for the moment and the train forced around by Annapolis. At Washington Eads was closeted at once with the Attorney General. The groundwork for a river navy had not been hard to lay, Bates said, for the President considered the Mississippi the key to the whole situation in the West, and had agreed that Captain Eads, who knew every mile of the stream was the one to devise gunboats for it.[9]

A Cabinet conference was called at once to hear Eads's recommendations. He put forth his plan for an armored river fleet with his usual hardly repressed enthusiasm. Metal-covered boats had long been used in commerce, he said, and the prejudice against them for war purposes was unfounded. He gave a description of what he thought the boats should be and an opinion of what they could do. They were absolutely essential, he believed, to recover the lost half of the river, without which no land victories could endure. Secretary of War Cameron thought that the whole notion of a river navy was absurd and chaffed it heartily, but the rest of the Cabinet endorsed the plan and ask James Eads to write out a description of the vessels he had in mind.

He shut himself up in his room with paper, pencils and the phantom gunboats that he had long ago shaped in his mind, and worked furiously. He described and sketched the boats, threw in some shore batteries to protect the streams still in Union hands, and sent the whole to Secretary of the Navy Gideon Welles. Presently a distress call reached him from Welles: Would Captain Eads come and explain the designs more fully to his aides? Nothing like these flat, shallow, metal-covered vessels had ever come to their attention before. The riverman, twinkling at the serious faces of the officers, could see a long, deep-hulled wooden ocean ship wedged firmly in every mind—the Navy had never sailed the Mississippi or been caught in close quarters, perhaps, between riverbank batteries.

He was put to it to make the unusual boats appear reasonable, then essential. He pictured the shallow streams, the fort-crested bluffs rising above them, the lack of seaway to escape bombardment, he used logic, eloquence, and likely a joke or so. The Navy Department adopted the ironclad boats unanimously and selected their designer to produce them. In order to hurry the program, it was suggested that he convert a few strong steamboats to ironclads. Captain John Rodgers, of the Navy, was placed in charge of the prospective fleet, and detailed to go with Eads to see the construction started.

Time, as always, was precious to James Eads. The sooner the gunboats were ready the sooner the Mississippi would be unshackled. He was tossing his clothes into valises when word came to him that Secretary of War Cameron, who had ridiculed the idea of a river navy and tried to defeat its adoption, now abruptly claimed control of it, the Army having, in a tussle of interests, gained sway over the inland rivers. The assignment from the Navy was nulled. Eads sat with his bulging bags and waited for the Army's orders. None came. Secretary Cameron, his point won, had faded from the scene. Eads's appeals to him fell into the silence that had closed about the river navy. At last, after "vexatious delays," the orders given to him and Rodgers by the Navy were repeated by the War Department. The two men hurried off.[10]

Almost from the beginning the ideas of the riverman and the ocean sailor clashed. At Cairo, Eads came upon his Submarine Number Seven, the once powerful twin-hulled snag boat *Benton*, the very thing with which to start the armored fleet. This strong craft, covered as a single hull with an iron casemate, would, he was convinced, make a superb gunboat. Captain Rodgers did not think so. Eads knew of several stout, light-draft steamboats on the Missouri River that would be excellent for their purpose, as the iron plating would not draw them too low in the water. Captain Rodgers, later to become a hero in the ocean navy, a rear admiral and a warm advocate of iron-plated vessels, was not now interested in gunboat armor—he had sailed the seas in various warships and none of them had been sheathed in metal. He was going back up the Ohio, he said, buy some boats he had seen, have their bows covered with oak timbers and their engines and boilers somewhat protected by coal bunkers. Eads dared a further word: in any case, the conversion work had better be done on the Mississippi, so as not to risk having the vessels tied up or grounded by low water on a tributary. Captain Rodgers brushed this aside—in all his years in the Navy he had never had a boat grounded. He left alone for Cincinnati, where he bought three side-wheel packets and began to alter them after his own design.[11]

James Eads, the river expert who had been called to Washington to devise a river fleet, found himself shelved. Full as he was of his plan to open the Mississippi, he knew that his civilian will must bow before the authority of the military officer in charge. But what could three makeshift wooden gunboats do to free the river? The enemy shore batteries would riddle them. Sore to his very soul, he could only report his situation to Washington and return home. On his way he made up his mind that the Mississippi would have an armored fleet. He would see that it did. Eunice, who had watched him leave so buoyantly on his mission, had never seen him as grim as he was on his return.

And grimly he tried to wait for instructions from the War Department, but time was drifting by. He turned to some of his friends about the stalemate, Frank Blair might be able to get under glazed Washington officialdom, he knew the Capital through and through—his father was editor of the *Globe* there, and the Blair mansion was in the very shadow of the White House,[12] and Washington well know Frank's head-on attacks and copious invective. Erastus Wells might help the cause of the ironclads, he had a way of getting things done. Attorney General Bates, Secretary of the Navy Welles and his assistant, Captain G. V. Fox, were doing what they could. Having pulled all political strings in reach, James Eads played chess ferociously and watched the growing turmoil around him. The Secessionists were pressing for control of Missouri, and if they won, the port of St. Louis and another long stretch of the Mississippi would be clipped from the Union. The country, already cut in two, north from south, would be quartered, east from west.

Still May of 1861, the hedges flowering, the sky deep blue between the trees. Usually at this season Eunice and the girls held bubbling conclaves about whether skirts would be narrower or sleeves fuller, the sewing room would foam with muslin and silk—six feminine wardrobes being replenished. To James Eads this flutter over gowns and bonnets was gay and diverting, as much a part of spring as the robin's song. He liked to see his family modishly turned out. May, and no pretty frivolity, only a silent waiting for they knew not what, or a tense anxiety when he was detained in town by drilling with the Home Guards. Sullen prophecy of an inevitable clash between factions hovered in the air.

On the evening of Friday, May 10, James Eads did not come home for supper. Real fear settled over his household. There had been a crackle of gunshots in the afternoon, not far away, and neighbors had brought word of a fracas at Camp Jackson, on the western outskirts of town, at Twenty-second and Olive streets. Governor Claiborne Jackson, the determined Secessionist, had stationed a force of the Missouri State Militia there, and it was feared by the Unionists that the camp was poised to sally out, seize the United States Arsenal at the river edge in the south part of town, and hold the city in a despotic sway. Captain Lyon, in charge of the Arsenal, had only a very small garrison to protect it. Frank Blair had wanted to do something about this, so he had managed to get some kind of authority to muster troops, had taken command of them and placed himself and his men under Lyon's orders.[13] Blair would not let the matter stagnate, a battle was inevitable. And whatever was going on, James was sure to be mixed up in it, Eunice knew. She peered anxiously through the windows into a downpour of rain that had begun at sunset.

It was late at night when James pulled up at the door, wet, mud-splashed and weary. Yes, there had been some shooting—the Arsenal forces had marched out, surrounded Camp Jackson and demanded its surrender. By two o'clock the camp had given up without a fight. But the three hundred captives had marched defiantly through the streets on their way to detention at the Arsenal, hurrahing for Jeff Davis, and southern sympathizers watching from sidewalks and roofs had jeered at the Union troops and finally fired into them. Shots had rattled back. Fifteen or twenty civilians and three Union soldiers were wounded. What had kept him so late? He had been on guard duty at the Arsenal.[14] Captain U. S. Grant happened to be loafing around there, down on a visit from Galena, Illinois, where he clerked in his brothers' leather store—he seemed less out of luck than usual. He had been about penniless when the county judges here turned down his application for county engineer, some time after he had resigned from the Regular Army. He was drilling volunteers at Galena, and hoped to get into action.[15]

James had no time for sleep, he was starting for Washington in a few hours. No one knew what would happen next, the Secessionists might attack the Arsenal, and with plenty of help from all over the state. The city and river needed more protection, there had been enough shilly-shally about it.

Early in the morning, Saturday, May 11, he took the ferry to

East St. Louis and boarded a Baltimore & Ohio train. At Lebanon, Illinois, on his way, he sent a long telegram to Secretary of War Cameron about the Camp Jackson capture, ending it: "Intense excitement in the city. Four thousand home guards under arms patrolling streets all night. . . . Left Arsenal at midnight, will arrive Monday at Washington. Our friends fear the return of Harney to St. Louis and protest against it." The Unionists had found their General Harney so dainty in his tactics that they suspected him of southern sympathy and resented him heartily.[16]

Even as James Eads wrote this, a fierce skirmish between the Home Guards and Secessionists was terrifying St. Louis. Six men, on both sides, were killed. Panic clutched the populace. On Sunday few persons ventured to church. Streets were deserted, doors bolted, curtains drawn. To make matters worse, General Harney did come back. The sight of the well-drilled, warriorlike German St. Louis citizens, many of them veterans of the 1848 revolution in Germany, who largely made up the detachments of soldiers that he sent to patrol the streets only added to the public nervousness—the protective troops looked far more menacing than the Secessionist mobs. Carriages and wagons filled to roll toward the ferries and across to Illinois, crowds surged to the levee to board steamboats for anywhere. After all, nothing dire happened.[17]

James Eads found Washington a bedlam, and no wonder, with so much needing to be done in so short a time. He saw the Secretary of War, gave him the details of the capture of Camp Jackson, the first aggressive move made by the Union anywhere in the country, and tackled him about the plight of St. Louis, of the river, of the northern states along it. They were in danger. No move had been made to recover the lower river. The ironclad navy—what had become of it? The Secretary appeared not to know. Eads wrote to Secretary of the Navy Welles, recommending Cairo, Illinois, at the juncture of the Ohio and Mississippi, as a naval base and outlined a plan for blockading, by use of gunboats, the commerce of the rebel river states.[18] At the Navy offices he found Welles and his assistant, Captain Fox, still eager for the ironclad fleet, but the matter was out of their hands.

Baffled and irate, Eads struck out about him, argued, pleaded. The very area in which President Lincoln thought that hard blows must be pounded was being ignored. Was the Capital unaware that the war affected the middle country? There was nothing more he could do here. He turned back home.

June, 1861, filled the Eads garden with roses and the lawn with callers. Unsmiling men talked in low, strained tones: Secession-minded Governor Claiborne Jackson had refused to obey President Lincoln's order to raise Missouri's quota of troops and the Convention had set him aside, assuming the reins of government. A hard-pressed government it would be. Some of Missouri's counties might throw off state authority, as Virginia's northwestern counties had, setting up a separate regime. Its lower border had already been invaded by Arkansas troops. Slave holding Kentucky, across the river from Missouri's toe, was likely to secede from the Union. The threat to the upper Mississippi was growing —why was nothing being done about a fleet for it? Something was, James Eads assured them, he was not letting the matter rest, and Frank Blair was right beside him in full, profane tilt.

July. The War Department had got around to the seemingly remote middle country and placed Major General John Charles Frémont, a noted pathfinder and son-in-law of Thomas Benton, over a newly established Western Department which generously included the whole legendary territory between the Mississippi and the Rockies. Frémont had entered upon his duties with a dramatic flourish and made himself felt by galloping recklessly through the streets with his bodyguard, his long curls flopping on his shoulders, dust or mud flying from their horses' hoofs.[19]

Still July, and the three wooden gunboats designed by Captain Rodgers, nearly completed and trying to reach Cairo in a low-water season, had grounded in the Ohio. When James Eads was angry it was with a wholehearted fury. There lay those boats, bleaching on a sand bar as he had feared they would, and the river's one pathetic hope of defense gone! While frustration ate into him, he heard that the Quartermaster General, in Washington, had advertised for bids to build seven ironclad gunboats for the western streams. The vessels departed somewhat from his design and they would not be quite fully armored, but they could do more for the river than a whole army. Eads made out his bid in tense absorption, pared it to the last margin of cost and pledged himself to complete all seven boats in the incredibly short time of sixty-five days.[20]

His daring proposal was accepted and, on August 7, 1861, in Washington, he signed the contract. These gunboats would be like large oblong shallow boxes, sitting six feet in the water, the sides and blunt ends sloping outward and down, the stern paddle wheel safe within the rear casemate, and each box hull pierced for thirteen guns. They would be creatures of the river itself, totally unlike the traditional warship.[21]

The building, or even conversion, of seven large boats in little more than two months would have been a large undertaking in normal times, it was a staggering task now. James Eads took swift stock of conditions over the country—mills, machine shops and foundries were closed, the workers dispersed. In a veritable tornado of energy he laid about him, tied up telegraph lines for hours at a time getting mills and shops in several states opened and manned, new ones hastily put up, machines for fashioning the armor plate built, materials shipped and foremen instructed. In two weeks he had four thousand men working in shifts seven days and nights a week, spurred on by tempting bonuses,[22] with William Nelson, builder of the Submarines, in charge.

In five weeks, on October 12, one of the gunboats was launched at the Eads shipyard (which he called the Union Ironworks) in Carondelet, a suburb at the south end of the city. Commodore Andrew Hull Foote, who had replaced Captain Rodgers in command of the Western Fleet, but under the control of the Army, named the vessel *St. Louis*. Foote had been skeptical and nervous about the unusual boats all along, predicting only the worst for them, but he now walked the new gunboat's deck with pride. Here, under his feet, was the strongest and least vulnerable war vessel that a Navy man had ever trod. Of this gunboat James Eads later wrote to President Lincoln, sending him a photograph of it: "The *St. Louis* was the first ironclad built in America. She was the first armored vessel against which the *fire of a hostile battery* was directed on this continent, and so far as I can ascertain, she was the first ironclad that ever engaged a naval force in the world."[23]

He was not satisfied with the time he had made, Eads confided to a few of his friends who trooped over the boat. Some of the War Department's designs had been impracticable and had had to be changed—and had anyone here ever tried to get a change made in anything once approved by that department? The labor shortage was more than hampering. And, besides, an eighth ironclad had been commenced, to build along with the others, a huge craft, to bear sixteen heavy guns. It would be far the best of the flotilla—the old Submarine Number Seven, the one-time snag boat *Benton*.

And what a row there had been over it! General Frémont had asked him if he knew of any river vessel that could readily be converted into an eighth gunboat, and he had trotted forth Number Seven, the one that Captain Rodgers had scornfully turned down. But since the firm of Eads and Nelson owned the vessel—which did not keep it from being superb for the purpose at hand—and fearing that his recommendation might be misconstrued as self-interest, he had asked Frémont to have the boat passed upon by impartial examiners, with orders to *under*appraise it. This had been carried out. The examiners were enthusiastic, Commodore Foote was delighted with the prospect of this large boat, to be completely armored, a thing the others had not been. But Quartermaster General Meigs had sniffed at the plan suspiciously. The boat belonged in part to Captain Eads! Something unusual must hang on that fact! And so it did—the extremely low price of $28,850. No one but a man in a ticklish position would have sold it at such a sacrifice. All this was explained to Meigs, but he still hinted at some obscure trickery. Eads had insisted upon further examination of the craft by different experts, all of whom agreed that it would make a splendid gunboat, and a cheap one. Meigs was tempering his stand now—he would tolerate the big boat, but he would never exonerate its builder. Anyway, old Number Seven, to be renamed *Benton,* was being converted, and no one had ever before seen such a warship as it would make.[24]

The four boats on the ways at St. Louis and three being constructed at Mound City, Illinois, not far from Cairo, were launched in quick succession, but the sixty-five days were ticking off fast. There was one grueling cause of delay that James Eads had not said much about, the government's failure to meet the payments that had fallen due. He had laid out enormous sums of money getting the gunboats started, and the default was growing until his whole fortune was about devoured. He was sacrificing his family's security in his resolution to recover the lost river. Over and over he reminded the War Department of the lapse, yet no payment came. He borrowed from banks until they would lend no more. On October 28, 1861, he revealed to Commodore Foote that the work would have to halt, all resources were at an end. Years later James Eads wrote of Foote: "He was, ordinarily, one of the most amiable-looking men; but when angered . . . his face impressed me as being most savage and demoniacal."[25]

The Commodore was angered now, his sharp black eyes burning with rage—he was a commander in danger of being without a fleet. He telegraphed the Quartermaster General that the gunboat construction could go no further without money. "What," he demanded frantically, "shall I do in the premises?"[26]

This cry fell into limbo, too. The government departments were swamped with demands and almost dry of funds. No payment came. James Eads, haggard but determined, made a plea to his friends: The war was six months old, the lower Mississippi was still cut off, the upper river threatened. The whole cause of the Union could be lost here in the middle states. Could they pool some money for the ironclads? They might never be repaid, he could promise them nothing. Hands reached for checks, gold clinked from hiding places. The hammering and riveting at the shipyard went on.

It took James Eads a hundred days, not sixty-five, to build the first seven ironclads and nearly complete the mammoth *Benton*. Late in November he saw four of the boats start downstream to Cairo.[27] Folk along the shores stared in amazement after the squat vessels that hugged the water, shouldering their bows pugnaciously. No one could have believed that river craft could embody such strength and ferocity.

In a few days more the *Benton* was ready to sail, and a more formidable warboat had never floated on any water. Shorn of its derricks, pumps and paddle wheels, its twin hulls planked together over their decks and under their keels, and their outer sides extended forward to form a single bow, it was entirely sheathed with armor, most of it three and a half inches thick. The iron casemate enclosed a single large paddle wheel, set forward from the stern. This boat was 200 feet long, and wider, with its 75-foot beam, than any warship on the seas. Foote, lately given the title of flag officer, had chosen it for his flagship. Even Quartermaster General Meigs had been won over to it and to its builder.[28]

When the leviathan snorted away from shore, James Eads was aboard. Foote, now at Cairo, had asked him to be on hand to give aid if the heavy vessel stranded on a shoal in the unseasonably low river. "It was December," Eads wrote of this trip, "and the water was falling rapidly.... Ice had just begun to float in the Mississippi when the *Benton* put out from my shipyard in Carondelet for the South. Some thirty or forty miles below St. Louis she grounded."

He had to stand helplessly by while the boat headed for trouble and while the naval officers, salt-water men, put out an anchor in the wrong direction to haul it off the bar. No one had informed the officers that he was moving about on deck in the icy wind because he had been placed in charge of the voyage down the shoaling stream, and not as an overanxious passenger who put in an unasked suggestion now and then when they were too busy to listen. He found himself without authority to prevent disaster, although he would be held responsible for the safe delivery of the vessel. Having learned in his tilts with Captain Rodgers that civilian disagreement with the military was futile, he watched the men work all night, getting the big hull "harder on the bar than ever."

Morning found the crew exhausted, the officers hopeless. The water was still falling, the massive ironclad settling deeper in the mud. Now, James Eads believed, the baffled officers would listen to him. He waded in, got authority to take charge, ordered the crew to bring out some eleven-inch hawsers that were being carried to the flotilla at Cairo, fasten them to a large tree ashore and haul in the opposite direction from that in which they had been pulling. Inch by inch the straining hawsers drew the wedged vessel from the shoal.[29]

Meanwhile, Flag Officer Foote, in a dudgeon over a message that the gunboat was in trouble, had sent a telegram to the Secretary of the Navy: "It is with deep regret that I inform the Department that the large and effective steamer *Benton* is badly aground seventy-five miles up the Mississippi. . . . The disaster is delaying our whole expedition." And he telegraphed Quartermaster Wise in St. Louis: "Send by return of mail contract with Eads."[30] One might suppose that the *Benton's* grounding had lasted for days while the rest of the flotilla waited with full steam up to go at the rebels. Actually, one night was lost, and the other gunboats, lacking men and supplies, were to tug at their moorings for many a week before an order for real action was given.

When the big ironclad steamed impressively into Cairo, the Flag Officer, his grievances having vapored, welcomed it proudly, asking all the Navy officers to look over it and meet its builder. He grew expansive and invited U. S. Grant, who had been abruptly promoted from colonel of a volunteer regiment to brigadier general, in July, and stationed there, to bring his officers aboard. The vessel was slow, Eads admitted—the engines that had hurled the former snagboat with ease against drift trees rooted in the river bed, snapping huge trunks as if they were twigs, labored under the weight of its armor and heavy guns, and would need augmenting. Nevertheless, Foote wrote the Quartermaster General: "The *Benton* is greatly superior to any gunboat I have ever seen. Every officer here pronounces her the best gunboat in the Union." . . . Upon close scrutiny he had increasing confidence in Mr. Eads, he said, but intended to watch all that he did.[31]

While Flag Officer Foote belabored the Navy and War Departments for officers and crews for his fleet, James Eads went to Washington to collect what he could of the money due on the gunboats. He was in sharp need of funds with which to ease his debts and to pay for alterations of the *Benton's* engines. He cooled his heels for four days before he could see Quartermaster General Meigs, and the interview was most depressing: The gunboats had not been completed on time, Meigs held, therefore a forfeit of ten thousand dollars on each of them was in force. As for the remainder of the three hundred thousand dollars' worth of accounts presented, for various reasons it could not be met either. Eads protested the forfeits, he had been delayed by lack of funds, by tardiness in government approval of essential changes in the gunboat designs, and by adding an eighth vessel. His words were wasted. He telegraphed Foote: "I can obtain no assurance of receiving a dollar and must return as I came."[32]

Back in Cairo, he found the air thick with gloom over the lack of crews. The Flag Officer, seething, had written the Navy Department on January 11 that none of the promised thousand or so men had arrived. Eleven days later he wired the department: "We will endeavor to make these boats efficient with six hundred men. . . . Will you send them?" The very next day he had wired again: "Can we have six hundred men?"[33] And still none came. Men, men, men, the Army, Navy and war factories were demanding—men who had been so cheap and unwanted hardly more than a year ago.

James Eads, having the *Benton's* engines strengthened, funds or no funds, was as impatient as Foote for the other gunboats to be manned and off on their task of opening the Mississippi. Why else had he toiled day and night, paid bonuses from his own pocket, borrowed, begged, loaded himself with interest? Everyone at the Navy base was on edge for action. At best, life was not exhilarating at Cairo, a drab port sprawling on a mud flat that was ever inviting a flood to roll over its levee, if only to wash out the dingy St. Charles Hotel, the only loafing place. The days spun out tediously.

St. Louis was bleak and gray, the piercing winds seemed to sear the very hearts of the people. Bitterness rose easily to the lips. Unionists were disturbed over the lack of protection—a state so torn with dissension might rend itself. February was at hand, the war was ten months old and the river still squeezed shut. Was nothing being done about it? The costly fleet, what had become of it? Except that several of the gunboats, led by the *St. Louis*, had poked here and there on the rivers, reconnoitering, scaring farm hands on shore half to death, and firing at the strong shore batteries at Columbus, Kentucky,[34] nothing had been heard of it. Had it sunk under the weight of its own armor? James Eads met these barbs on every hand, he faced the banks and his friends who had helped him finance the ironclads and admitted that he could collect little from Washington. As for the gunboats, they would go about their work presently when crews for them had been pieced together of farmers, factory workers and loafers, most of whom had never set foot on a deck before.

On February 6, 1862, the news was shouted over St. Louis that the river fleet had gone into real action, it had scored a victory! Flag Officer Foote, in a temporary flagship, the *Cincinnati*, with three other ironclads and the three wooden gunboats, manned by mixed but loyal crews, had crept into the Tennessee River, fishing up enemy mines as they went, and headed for Confederate Fort Henry. This stronghold, lying on a marshy east bank, menaced the movement of Union troops southward and had to be reduced. At dawn, with river and shore merged in a slate-gray fog, the *Cincinnati* had moved out from behind Panther Island, leading the three other ironclads abreast, the wooden boats lagging behind. The sun had come out suddenly and the enemy lookouts, startled by the strange war vessels almost upon them, gave an alarm. The gunboats poured destruction into the fort from behind a cover of smoke. A Confederate officer reported that the onslaught "exceeded in terror anything that the imagination had pictured of shot and shell, plowing roads through the earthworks and sandbags, dismantling guns . . . setting on fire and bringing down buildings within the fortification, and cutting in two, as with a scythe, large trees in the neghborhood." The fort, desperately defended, gave up in a few hours to General Grant, whose troops had been landed nearby from transports.[35]

Just a week later good news again flashed up the valley. The Union gunboats had turned back into the Ohio and up the Cumberland River. There on one of the hills that rose sharply from the water sat Fort Donelson, with ravines and entrenchments reaching widely around it and three tiers of batteries rising from the river to its bastions—a tough nut for land forces to crack, and considered impregnable from the water. On February 13, 1862, Flag Officer Foote, aboard the *Cincinnati,* led three ironclads and the trailing wooden vessels and set up a thunderous shelling. At the battle's height Confederate General Pillow, it was said, telegraphed the Governor of Tennessee: "The Federal gunboats are destroying us." The gunboats took a fierce pounding, too, suffering a hundred and forty-one wounds. The man casualties were low, but among them was the Flag Officer, who received two wounds, the more serious one in a foot. He made little of it. He had forced the surrender of the splendid fort, and the Union Army was moving into it.[36]

Two hard blows had been struck to free the Mississippi. The armored gunboats, still unpaid for and, for the most part, the property of James Eads and his friends, had given the North its first important victories. The financial pinch on Compton Hill felt less cramping, creditors were easier to face. The pall of pessimism over the whole country lightened. Accounts of the ironclad fleet veered to the sensational—the grotesque boats were horrifying, deadly. Nothing like them had ever been known before. Men who stamped in out of the cold to back around the stove in Eads's shipyard office swore that St. Louis must celebrate, hang out flags, fire its cannon. Just one cannon shot, they were reminded, would empty the streets and bar every door in the city. Directly a parade was being planned, the town bubbling with preparations—many a hatchet was buried, to be dug up later. On February 22, 1862, a few cannon did boom, but it was across the music of bands marching in a procession that took two hours to pass each cheering crowd on streets or rooftops.[37]

James Eads was likely tired when the day was over, for he had yet to learn to take adulation without embarrassment. And he was thoughtful. He had written Commodore Foote that he rejoiced "with the prideful pleasure of the poor armorer who forged the sword that in gallant hands struck down the foe."[38] But he was not fully satisfied with the gunboats he had built, he could confide to Eunice—their manipulation was awkward, their gunfire slow, the armor of all of them but the *Benton* was incomplete. The river needed stouter, faster, more rapid-firing ironclads, and he was going after Washington about it. Eunice, torn between pride in her husband and fear for his health, could only wonder at him and tremble. He ached with fatigue, but his energy drove him on. There he sat, planning to crush himself under another nearly superhuman task, and filled with contentment over it.

Always impatient to reach a goal, James Eads felt that time was frittering by. The war was nearly a year old, and the Mississippi still cut in half. Yet, the ironclads had had another triumph. On the very day, February 22, 1862, that flags had fluttered and bands played to celebrate their earlier victories, the gunboats had bombarded Columbus, Kentucky, perched on high bluffs above the Mississippi, until its doughty General Leonidas Polk, an Episcopal bishop, and his twenty thousand men surrendered. Now Commodore Foote was heading six ironclads and a wooden gunboat down toward Island Number Ten, at the border of Ten-

THE FIRST IRONCLAD GUNBOATS 69

nessee.[39] A remarkable commander, James Eads considered him, he should have more and better boats.

Eads had not spared the military departments in Washington repeated airings of his blunt appraisal of the fleet, and they may have taken effect, for in April, 1862, he had a telegram from Secretary of the Navy Welles to report to him for a conference.[40]

The Capital was sunny, beautiful and grave. The war had already lasted longer than was generally predicted, and yet the nation was still making a steep uphill struggle just to prepare for war. The Navy wanted two fully armored gunboats built, Welles revealed. Only two? That was disappointing. These boats, the Secretary went on, must be of very light draft, as they were for use on the tributaries. Oh! For the tributaries—but the big river needed boats, too. Eads's arguments were brushed aside.

Shortly he presented plans for a vessel which, in spite of its plating, had a draft of only five feet, a real achievement. Captain Fox, Assistant Secretary of the Navy, shook his head over it. "We want vessels much lighter than that."

"But you want them to carry a certain thickness of iron," Eads protested.

"Yes . . . to be plated and heavily plated—but they must be of lighter draft."[41]

James Eads picked up his plans. He was asked to do the impossible—all right, he would. His hotel room swallowed him. When he came forth again he had a design for a boat whose edges were only a *half foot* above the water, the deck curving upward four feet to the center. It was on the order of the partly submerged *Monitor* lately designed by John Ericsson, a versatile Swedish-American inventor, for the United States Navy; but its draft was only three and a half feet, eight feet less than that of the *Monitor*.

His rotating gun turret, Eads explained eagerly to Welles and Fox, was on a plan entirely different from any other. The guns would be loaded below deck and raised by steam—they could be fired every forty-five seconds, seven times as fast as the guns in John Ericsson's turret, and they would have greater elevation. The portholes would be so small that enemy shot could not enter them when the gun was firing and they would close automatically the instant the gun muzzle was withdrawn. The recoil, which ordinarily was five to seven feet on ten-inch guns, would be taken on a cushion of steam, cutting down gun strain. It would take only three men to perform the whole operation, and—[42]

But Welles was satisfied with the *Monitor* guns, they had held the Confederate *Merrimac* to a draw only a few weeks ago, March 9, a virtual victory. To James Eads's keen disappointment, his turret was rejected, but he was permitted to modify the Ericsson turret by bending down the edges of the floor so as to give a better gun range.

He had hardly returned to St. Louis and settled to his drawing board than he was on his way back to Washington, in considerable excitement. Secretary Welles wanted four more monitors, large ones, each with two gun turrets, for the lower Mississippi. One can imagine that Eads's pencil was busy all the way to the Capital.

When James Eads showed Secretary Welles his sketch for the large monitors, conspicuous on it was his rotating steam turret with the fast-moving guns of nearly unstinted range. Welles was annoyed, that turret again! He had ruled it out once, had he not? There was a clash of wills, a compromise: Eads could place one of his turrets on each of two of the four boats, provided he would change it at his own expense—about thirty-five thousand dollars— to an Ericsson turret if it proved unsatisfactory. But the other six turrets on the four vessels must be of Ericsson's design. In any case, Welles said, the sketch had better be shown to Flag Officer Foote for approval.[43]

It took less than a second for James Eads to jump at this suggestion, for a conference with Foote meant a visit to the flotilla which now sat a few miles above Island Number Ten, shelling the shore batteries of the enemy and awaiting a chance to worm down the narrow channel past the murderous fire from the fort so as to trap the island between forces above and below it.

At Cairo, Eads got on a small tug and headed downriver. The muffled roar of cannon reached him long before the acrid smell of powder tainted the wind. Writhing smoke hid the shore line, the tug crept cautiously. Shouted orders sounded nearer, belches of fire limned, for separate instants, the ironclads crouching on the water, ferociously at work. The monster yonder was the *Benton*. The tug made for it, Eads climbed aboard.

His spirits were soon dampened. Flag Officer Foote was in no mood for a conference, even about gunboats. He was suffering greatly from the foot wound received at Fort Donelson, and annoyed over many things: Island Number Ten, heavily fortified, was cut off from the shore by swamps, not navigable water, and the single strip of river past its front was narrow and swift. How could gunboats fighting downstream in such a current be kept in position? If disabled by the shore batteries they could not float away to safety as in the upstream attacks upon Forts Henry and Donelson—a man used to plenty of sea room felt cramped in this river war. Some of his guns were defective, made in 1855 and rejected then for want of strength. Much of the ammunition dated back to the Mexican War and exploded too soon, or never. There were the accursed mortar boats that had been tacked to his flotilla without his leave, vile nuisances—he "awoke every morning with a mortar boat on his heart." Besides all this, there was the supreme indignity, Army control! General Halleck's orders held him back when he should go ahead, and the other way around.[44]

An officer stepped up with a package of mail for the Flag Officer, who selected a letter, read a few lines of it and said calmly: "Mr. Eads, I must ask you to excuse me for a few minutes while I go to my cabin. This letter brings me the news of the death of my son, about thirteen years old."[45]

After fifteen minutes he returned, quite composed. Eads tried to divert him, and even managed to bring a wan smile to his lips, then asked him if he could be assigned a place on the *Benton* for the night—he hoped to make the most of this visit, going from one gunboat to another to see them in action. Foote became grave: "I cannot permit you to stay here a moment after the tug is ready to return. . . . You must not stay." Every second was full of danger, and a shipbuilder was too essential to risk idly. Eads was minded to argue, but a shell that struck the water only fifty feet away spoiled any plea he could make.[46] He found himself being conducted unceremoniously back to the tug.

In his shipyard office he worked out the details of the two small monitors, impatient to get at the large ones, and figured how to stretch his funds to cover urgent bills. A belated payment on the *Benton* had lightened his debts and furnished a donation for Flag Officer Foote to use in relief of the suffering that the gunboat victories inevitably caused. Spring dusk would have closed in by the time he started home from his day's work, the sunset copper already grayed from the river. The thud of the horses' hoofs on the unpaved suburban streets and the jingle of harness were soothing, they helped to push out shipyard anxieties that sought to steal a ride home with him.

The hoofbeats swung around the curved drive and stopped at the side door. James Eads climbed the side stairs to the second floor. When he came down the front staircase later he was no longer the hard-driven shipbuilder, but a meticulously dressed, apparently carefree gentleman. He held himself so erect that one hardly noticed how flat his chest had grown, his smile belied the lines of strain in his lean face. Some of his guests had likely arrived already, neighbors, business associates, Army or Navy officers, young friends of his daughters. Others were alighting from their carriages at the side door, climbing the side stairs to the second floor, laying aside their wraps and making a grand entrance down the impressive main stairway to be ushered through the reception parlor into the long drawing room.[47] They thronged presently to the ample dining room. A supper was not what it used to be before the war shut off the flow of food, but the host kept the meal gay, if ceremonious. "His manners," one of his grandsons wrote, "were rather those one expects in a European gentleman of leisure and high breeding, than those of a man who had worked hard all his life."[48]

Yet he was without pretense and talked quite simply about his experiences as chore boy or steamboat clerk. He believed fervently that one should respect the trade that supported him, and found the spectacle of young Americans being "reared in the belief that labor and respectability were incompatible" grotesque and sad. "Labor," he said, "when coupled with knowledge, becomes a mighty engine of power."[49] Earnest work was part of his creed.

Still spent from the day's demands, he would relax on the lawn, his guests about him. Distantly came the landing whistle of a steamboat—Eads dared not let himself wonder if it brought iron for the gunboats. The garden was pale silver, the lights of the city twinkled through the mist. The night's beauty stirred him. In a moment's silence he would murmur a quotation from a favorite poet, and then he was off reciting verse after verse of Tennyson, Whittier or Poe.[50] However casually the lines began, they were likely to grow in fervor until they rang. New visitors would marvel indulgently at this enigmatic man, salvage diver and boatbuilder, dapper and dignified, simple and unself-conscious, as, with a hunger for rhythm, he declaimed the poems of the romanticists.

The ironclads were never long out of the forefront of their builder's mind. How could the new ones be hastened? What were the old ones doing to rescue the river? News good and bad reached him from the Mississippi fleet: The black, stormy night of April 4, 1862, had given Commodore Foote an opportunity to slip the ironclad *Pittsburg* down the narrow passage past Island Number Ten, the *Carondelet* followed it. Transports with men and supplies were sent through a canal that had been secretly cut by Union soldiers and sailors across the neck of swamp forest far behind the island. Gathered below New Madrid, Missouri, which, although downstream, was several miles north of Number Ten, owing to a sharp horseshoe curve of the river, these forces formed one jaw of the long-planned trap. The defenders of Number Ten, realizing that they were about to be swallowed, sent a flag of truce, on April 7, aboard the waiting *Benton*. Foote and the fleet had then dropped down to Fort Pillow, on a bluff about eighty miles upstream from Memphis. Below Memphis waited an enemy flotilla.[51]

For a short while there was a dearth of news about the ironclads. Gentlemen who sat on the Eads lawn, sipping iced wine, talked more about "greenbacks" than gunboats. These treasury notes, issued by the government, had actually been declared legal tender —it was enough to make "Hard Money" Tom Benton turn over in his grave. Commodore Foote's wound had grown serious and he had left his fleet—Captain Charles H. Davis had succeeded him. What was all this talk about the Missouri Radicals wanting a new Convention that would end slavery in the state? President Lincoln would attend to that directly, he was hinting that he might proclaim all slaves free, in both North and South, if the rebel states did not lay down their arms. He had a plan to colonize the freed Negroes in Central America, perhaps Nicaragua, and Frank Blair was about to get an appropriation through Congress for this purpose. Blair was having quite a feud with his cousin, Gratz Brown, who had deserted the ranks of the Liberal Republicans for the Radicals and was advocating all sorts of addlepated notions, among them, universal suffrage, "even for Negroes and women!" All Radicals were "damned whelps," Blair said, blessed with more sentiment than sense.[52]

But the most absorbing topic, discussed with mixed feelings, was the sea fleet that Commodore Farragut had brought from the Gulf into the Mississippi, warships, sixteen deep-hulled wooden vessels, some of them sailing ships, and more than twenty small mortar-bearing schooners. Ocean battleships on the Mississippi! Farragut had had great difficulty in getting the fleet over the Gulf bar outside the delta, it had taken two weeks for tugs to inch the *Pensacola* across the rim of mud, the *Colorado* could not be pulled in at all. James Eads was likely to grow restless at the mention of the Gulf bar, this obstruction had challenged him for a long time but there was nothing that his river-clearing boats could do about it.[53]

Steaming or towed, Farragut's ships had reached a tree-grown bend not far below Forts St. Philip and Jackson that guarded the river about thirty miles above the delta head. Here, their masts disguised with tree branches, they had lain several days, throwing shells into the forts on computed aim, being shelled, and staving off fire rafts sent down against them. They had run past the forts, fought a Confederate flotilla above, and reached New Orleans, which surrendered to them. Farragut had found the waterfront there ablaze under a gray sky, cotton had been carried out of the warehouses and torches put to it on the wharves, sugar, rice and molasses had been poured out and riffraff were busily scooping up what they could. The white lines of steamboats were blackening in their own flames. General Benjamin F. Butler, docking his troopships there soon afterwards, was met by a furious mob, the stench of his reputation having wafted ahead of him. The General had slunk ashore, ungainly and sullen, keeping well in the background. But he was swaggering now, subjecting the populace to every indignity, insulting, looting them, nauseating them with his own personal immorality. Comment upon this rose high above the clink of sangaree glasses—New Orleans might be an enemy for the moment, but she was a sister port to St. Louis, with many strong ties of blood and custom. Butler's treatment of her citizens was outrageous.[54]

Had Captain Eads heard about Colonel Charles Ellet's six naval rams, painted black for "frightfulness," which had joined the ironclads? It would take a fellow like Ellet, brilliant and bullheaded, to invent a warboat that attacked enemy craft by running head-on into them! More than twenty years ago he had proposed to throw a suspension bridge across the river from St. Louis, as though anyone could bridge the Mississippi here where it raced past like a galloping herd! For many months past, now, Ellet had been nagging the government for permission to build the rams, but got it only after the South had begun using his idea. There was a Confederate ram below Memphis by the time he got his own out on the river.[55] When were the ironclads going to tackle the enemy fleet? What was the delay? James Eads did not know, he could only wonder, impatience seething in him. There they sat above Memphis, alert for a word from the Army! At such a moment he found it safer to recite poetry than to comment.

In the second week of May bulletins of the first real naval battle of the ironclads—that is, a struggle against boats, not shore batteries—were hurried to James Eads by many an excited messenger. On May 10, there had been an attack on the armored river fleet by the Confederate flotilla near Fort Pillow. It was a pitiless conflict, the ironclad *Cincinnati* was sunk,[56] but the enemy had got the worst of the fight and had limped, badly disabled, to the shelter of the fort. This was great news for the ironclads' builder, for it bore out his belief that, given a chance, his gunboats would clear the enemy out of the river. It was pleasant news to take home to Eunice, she beamed upon his successes.

Three weeks later, Fort Pillow was abandoned by the Confederates, and the rebel gunboats dropped downstream. The ironclads had overtaken them and there was another desperate engagement. Ellet's rams had done much to disconcert the enemy, but the "old war-horse" *Benton* had played the leading role, pursuing the fleeing boats with "an accuracy of fire and execution really terrible."[57] Such a boat! Shells did not bother it, enemy rams fairly bounced off of it. This blow to southern naval power was final, James Eads believed. The way would be clear now for the ironclads to attack Vicksburg, which sat on a high crest above the river, guarded by four miles of shore batteries that rose in tiers up the face of lofty bluffs. When that Gibraltar fell, the Mississippi would soon be free.

CHAPTER VI

THE RIVER MONITORS

LATE June of 1862 rode into St. Louis on a soft wind from the plains. The midday sun painted the brick houses and brick streets a light vermilion and whitened the now dingy steamboats at the wharves. It was hot in the shipyard, the river was glassy and blinding. James Eads, moving about among the too few workmen on the ways, found them listless, disheartened. There were boats arriving with troops or cannon, there were barges of coal or rafts of lumber, but no craft with iron plates or engine parts, they complained. Their hands were always reaching for something that was not there—they could not make gunboats out of promises and thin air. Back at his desk, with the heat throbbing in his temples, Eads sent his threadbare appeals to eastern mills to hurry material to him as best they could. He wrote everywhere for men, skilled or unskilled, he hired whatever straggler happened along, then turned back to his drawing board and his plans for the large monitors.

Already demands for the hardly commenced gunboats had become insistent, the most impatient of them from the officers of the older ironclads. They needed the monitors for a sweep they meant to make down the Mississippi, if they could ever get the chance—Army orders held them back as mere appendages to the land forces, frustration walked their decks and manned their wheels. Farragut was actually winding up the Mississippi with his ocean vessels, and the river gunboats pinned to the shore! Farragut making for Vicksburg, as if he could take it with ships! When would the new gunboats be finished? It was not ships the river needed.

Farragut, sitting just below Vicksburg, was no better pleased with things as they were. He had sent an explosive note to the Navy Department saying that the difficulties of the river had proved more destructive to his vessels than enemy shells. He dropped downstream lest he be caught in low water, while landsmen jeered his tall masts and unwieldy hulls. Presently he was up near Vicksburg again, "by peremptory order of the Department and the President." And, under orders, he brought his fleet past the tiered fort, firing and fired upon, suffering much damage, inflicting none. "It seems to me that any man of common sense would know that this place cannot be taken by ships," he wrote in exasperation.[1]

James Eads, piqued that the ironclads were held at anchor while ocean vessels dominated the river, kept watch on a movement in Congress to lift the control of the Mississippi ironclads from the Army and place it with the Navy. Late in June he was cheered by the news that the move had succeeded, and that presently the gunboats could go on their mission of opening the river.

Before the new control had gone into effect they were started downstream, and on July 1, 1862, the two Union flotillas met. To most of the ironclad crews the towering masts and huge rearing hulls of the ships were a prodigious sight, to the salt-water sailors the crouching armored craft were no more than terrific floating batteries. Farragut dismissed the ironclads as "curious looking things . . . like great turtles." He pointedly did not like them. Or the rivers. The water was falling and he feared being stuck in Mississippi mud. After an indecisive skirmish with a Confederate flotilla that dared out from the protection of the fort, and witnessing the effectiveness of the "turtles," Farragut dropped downstream, orders or no orders.[2] Vicksburg remained untouched.

Work at the Carondelet shipyard was pushed feverishly, demands for the monitors pouring in on James Eads from every side. Ironclads, more ironclads, urged Flag Officer Davis. Ironclads, insisted Lieutenant Commander Seth Phelps in the Tennessee.

Ironclads, wailed Commander Porter in the lower Mississippi. Negroes toiling in the cotton rows or sugar cane now crooned a wishful chant, "Wen de Linkum gunboats come." Even Farragut, who had scorned the armored boats fervently, was to send a piqued request for them: "If I could see any great importance these vessels would be to Porter, I would not ask for them."[3]

The heckled shipbuilder combed the country for men and pleaded for carriers to bear material to him. By autumn the rivers were picked nearly clean of craft for civilian use. At midwinter of '62-'63, James Eads wrote to the War Department in some desperation: "I have repeatedly suffered serious disappointment in receiving my supplies . . . by the impressment of transports on the Ohio River. . . . I have now to rely almost solely upon the packet *Bostona* . . . and pray that some order be given which will prevent her from being taken also." To his lean comfort, the order was given.[4]

Other boatbuilders suffered like vexations. The strain put upon men and machines by the insatiable demands of war hovered, an impartial ogre, over the conflict. A Southerner, John Shirley, was called before a Confederate naval committee in February of 1863 and his loyalty sternly questioned because his construction of armored rams lagged. In his own defense Shirley testified that he had rebuilt sawmills, improvised machinery, sent everywhere for laborers, and "made all the efforts in the power of any man."[5]

Severe cold weather appears to have halted some of the work on the St. Louis ironclads early in 1863. Eunice Eads, for her part, could not be too sorry over this. Perhaps James, who had not been well, would rest a little. And he did, after a fashion, packing case after case of catawba grape wine to send to various friends. He was an inveterate giver of presents, and throughout the past busy summer had found time to get off packages of fruit and wine to ailing Andrew Foote, now a rear admiral, and had had a portrait of the hero painted and sent to him. Now, presently, came a case of Moselle wine to Compton Hill from Baron Gerolt, Prussia's chargé d'affaires, and a pair of diamond studs that Congressman Henry T. Blow, a near neighbor, had dispatched to "a diamond of the first water." And Edward Bates was having a cane made for him out of "heart of oak" cut from the sunken *Merrimac*. There were notes to be written, mock accusations to be made, facetious denials to be laughed over.[6]

The weather eased, James Eads hurried the work again at the shipyard. Need of the gunboats loomed greater than ever. The eyes of both North and South had fastened upon Vicksburg. There it sat on its terraced bluffs, flanked by guarding hills and ravines, impregnable, holding the river in its firm grip while fleets battled below it and armies marked time behind it. When he was putting the last paint on the light-draft monitors, Eads heard that General Grant and a strong force, convoyed by some of the gunboats, had crossed the Mississippi, marched down a piece and recrossed, so that Union flotillas and Union armies sat about the stronghold in a close siege. A battle might develop at any minute—ironclads, more ironclads.

On a June day he headed the first monitors he had undertaken to build, the light-draft *Osage* and *Neosho*, downstream. Never on this river of varied craft had anything more astonishing than these boats been set afloat. The iron-covered decks rounded up to their centers almost from the water, the turrets stood only a few feet higher.[7] Admiral Farragut would find these gunboats, with the river slashing at their curved backs, more than ever like huge turtles.

The whole nation watched the siege of Vicksburg. On July 4, 1863, the very day after the triumph of the eastern Union forces at Gettysburg, the famished river Gibraltar surrendered. The news flashed up the valley, men carried it inland on horseback, drivers of ox teams shouted it to each other on backwoods roads: The Mississippi was open! All the way down! Many made their way to Compton Hill to congratulate James Eads on the part that he and his ironclads had played in the victory, and to deplore that it had not come a few days earlier—what a Fourth of July they would have planned!

On the following day, July 5, Eads launched the large monitor

Winnebago. He was extremely proud of this boat. A Navy inspector, J. W. King, after examining the Eads turret on it, sent an enthusiastic report to Secretary Welles. This turret, he said, differed "in all respects from any heretofore constructed." Unlike the Ericsson model, it did not rest upon the deck and revolve about a central shaft, but extended, armored for some distance, below deck and revolved on spheres supported upon a circular box beam secured to the bottom of the vessel. The machinery performed a combination of functions with knowing ease that astonished him, revolving, reversing and regulating the motion of the turret; raising and lowering the platform containing the guns; moving the guns to firing positions and back again, reducing and cushioning the recoil, opening and closing the port. It was the first time in the history of artillery practice that heavy guns were manipulated entirely by steam.

"The design, construction and arrangement of the details of the machinery is highly creditable to the ingenuity, mechanical skill and ability of the inventor . . . who had to contend with all the disadvantages common to a light draft vessel," King wrote.[8] Whether permission was given James Eads before or after this report, he finally had placed one of his turrets on each of three of the four large monitors, and had redesigned one of the Ericsson turrets on the fourth.[9]

The Mississippi was open, James Eads exulted, yet it was under constant threat by the strong Confederate flotilla sheltered in Mobile Bay. Rear Admiral Farragut, on edge to attack, and frustrated by tedious repairs he had to make on what was left of his crippled fleet, was counting heavily upon the large monitors to reinforce him. Eads realized that these supergunboats were the pivot on which might turn victory or defeat in the Gulf, and that with the country all but wrung dry of men and carriers, it would take almost superhuman effort to complete them. The Carondelet shipyard was a treadmill of toil. The late summer heat beat down on the shed roofs, heat steamed up from the river. Men, iron, credit—the huge hulks on the ways were famished for them.

Their builder drove into the city and held serious conferences with bankers and steamboat officials, he combed the wharves and markets for workmen.

The streets were busier than they had ever been. Soldiers marched crisply toward the wharves, breaking ranks around a line of supply wagons. People pushed each other to board one of Erastus Wells's jammed mule cars, people crowded into the stores to spend their money on whatever they could find. The elation over the fall of Vicksburg had died down, a dull depression lay over all. The causes of irritation were many: the lack of sugar and molasses, of cotton and leather, and above all, of "hard change." The "shin-plaster" paper currency that had replaced the vanished small coins was an endless nuisance—a man liked money that clinked, that did not stick to his fingers or rumple in his pockets. And the burden of taxes—men congested the bank doorways while they argued that the war taxes were squeezing the life out of the country, and the government should issue more bonds. Bonds! others roared back, bonds to crush posterity under a mountain of debt![10] Bonds, income tax, sales tax, debt, inflation—why did not the government do something? Prices were running away, a calico dress cost a small fortune, boots would go to sixty dollars a pair!

Crippled or blinded veterans furnished a somber note in the streets, as did women in dull black with crape veils hanging from their bonnets. Even more distressing were the homeless, wandering refugees who trickled incessantly into town from the invaded southern territory, sent north by the Union generals because they had become "a serious impediment to military movements." They landed from boats, herding together at the wharves; they came in wagons and hastily fashioned carts, they plodded afoot, their worldly goods in bundles slung over their shoulders, babies in the arms of stumbling women. Nothing, it seemed to James Eads, had ever pulled at his sympathy like this human flotsam. It had been streaming here for months. "The greatness of their numbers appalled us," one St. Louisan wrote of the refugees.[11] The burden of providing for them had, at first, been thrown by the War Department upon local southern sympathizers, but James Eads had protested against this, reminding the military authorities that the war was "an accursed contest between brothers." He had put a check for a thousand dollars in his letter to start a fund for the homeless. After that, a Sanitary Commission had been organized to provide refugee care. This care was meager enough, funds were always running out, and the tragedy of the haggard newcomers weighed upon the heart of the tired boatbuilder as he went his rounds in quest of money or credit.[12]

Late in 1863 the four big monitors were nearing completion, despite lacks and delays. A few months more, and they would be out of the way, leaving room for a small flotilla of lightly armored boats, "tinclads," for patrol duty, and four large mortar boats. Fatigue pulled at James Eads. He fought it down, forgot his anxieties a brief while in Melmoth's three-volume *Cicero* sent him by Edward Bates, and went with Eunice, in December, for a short visit at the Attorney General's home in Washington.[13] Then he was back at the shipyard, and there he stayed until the new year had opened. The monitors must be rushed, it was January, 1864, and the mouth of the Mississippi under hourly threat from the enemy's Gulf fleet. He must rush something, Eads felt vaguely on a day when he could not rise from his bed. Doctors bent over the worn man gravely. He was not yet forty-four years old, but he was through. Specialists who were summoned gave the same verdict: James Eads, crushed under strain, would likely never work again.[14]

The winter winds drifted snow over the garden, shook the house and moaned at the windows. The exhausted man on the bed did not hear it. He was away under hot blue skies on the tawny river with his new ironclads . . . their turrets whirled, the guns moved up and down, faster, faster, the river blurred . . . iron, iron, no boats to bring it . . . the shed was hot and still . . . hot. The fever burned out, the spent man wrestled to piece together bits of memory: the gunboats . . . the *Chickasaw* was not finished, the river was waiting for it. Determinedly, with what patience he could muster, James Eads fought back his weakness, strove for recovery, grasped it, clung to it. Almost by sheer will he pulled up from his bed, learned as a child to steady himself and walk about the room. And the very first day he was able, he rode through the sharp, chill air to the shipyard. Not averse to sympathy or to dramatizing himself at moments, he seems to have made much of his illness in letters to his friends. "I am deeply grieved," Edward Bates replied, "that you have a painful and slow disease, though, as I suppose, not at all implicating life."

Still weak and tense, James Eads was badly shaken by a tragedy that attended the launching of the *Chickasaw* on February 10, 1864, which had been planned as a gay affair. As "the immense iron structure slid from the ways and plunged into the river, rising and floating like a cork," the laughter and applause of those watching it was cut across by screams of alarm. The anchor had gone overboard, the huge coils of its cable were sweeping the deck before them. Eads's oldest stepdaughter, Genevieve, and several others standing at the bow, were thrown into the water. They clung to floating timbers until rescue boats reached them and drew them, half frozen, aboard—all but one, a Mrs. Bradley, who had gone down in the icy water.[15] The sudden fright, the grief of the drowned woman's husband, left James Eads less well. His doctors insisted that he get away for a rest.

He was lured into a promise that he would go to Europe if Eunice and the girls would go with him. But he first wanted the Navy Department to inspect the new ironclads—"a mere gratification," the impatient Farragut testily labeled this.[16] And there was a Grand Fair in preparation in St. Louis to raise money and clothing for the now bankrupt Santary Commission's care of the refugees who still came in a thin but endless flow.

Appeals had been sent afar for donations to the Fair, money and gifts were coming from states east and west, and from abroad. James Eads, interested in any move that would ease the plight of the homeless war victims, soon found himself on the sidelines of a diverting controversy that had erupted right from the floor of the large building on Twelfth Street that was housing the occasion: The several breweries in the city had given a great quantity of lager beer to be sold, and the ever-alert temperance societies had pounced upon it in voluble protest. The skirmish was sharp, but when the Fair opened in May, the lager beer, sitting cheek by jowl with clothing, jewelry and objets d'art, was the most popular commodity. The Fair, enlivened by dramatic touches of various kinds, one of them the Old Woman in a Shoe impersonated by General Grant's daughter, Nellie, was a marked social success, it erased many antagonisms for the moment, and it brought in a half million dollars.[17]

Trunks had been set out in the Eads home, dresses of various sizes, cloaks and bonnets were piled about—a pretty nosegay of femininity the tired shipbuilder would have fluttering about him on his voyage, and there was nothing he liked better. Letters were coming from friends everywhere wishing him a pleasant trip and restored health. The party set off on the ferry about the middle of June, and stopped in Washington. While Eads was there, the Navy Department appointed him a special agent to visit the navy yards of Europe and acquire such information about gunboat types as might be useful to the Navy of the United States. He now held eight patents for gun turrets, ordnance operation, ship-o'-war pilothouse and special propeller,[18] and the military leaned heavily upon him to show the way to a new ocean navy.

Everywhere he went in Europe, James Eads was given a hearty social and official welcome, his most cordial reception being by Count (later Prince) Bismarck.[19] But his biggest excitement was news of the war at home, especially that about the part his gunboats were playing. He was particularly anxious to hear how those sent to Farragut in the Gulf would comport themselves.

Admiral Farragut, still making ready for the long-planned attack upon Mobile Bay, welcomed the large monitors and several of the older gunboats to his Gulf base. "Only give me the ironclads built by Mr. Eads," he had implored Secretary Welles, "and I will find out how far Providence is with us."[20] Providence and a strong fleet would both have to be with him, for the bay, a long, narrow arm of the Gulf, was protected by three forts, and harbored among the enemy craft the formidable armored and armed ram *Tennessee.* There might be submarines lurking there, too—the Confederate submarine *Hunley,* built at Mobile, an absurd vessel only five feet high and four feet wide inside, hand propelled by eight men, had, on February 17, torpedoed the U.S.S. *Housatonic* off Charleston, sinking with its prey.[21]

It was not until late summer of 1864 that Farragut dared the onslaught. Word of its outcome, sped to Eads abroad, repaid the builder of the monitors for many a weary hour. The *Chickasaw* and the *Winnebago,* with warships and mortar schooners, had steamed into the bay on August 4, their smoke, carried ahead by a light wind, screening them from the forts. The Confederate gunboats emerged from behind the forts, vomiting fire into the smoke, the land batteries opened up. The *Chickasaw* boldly singled out the enemy ram *Tennessee* that bristled with guns, and swung in behind it. No matter which way the ram veered, there was the alert monitor "holding on like a bulldog" to its stern, clattering shells at a speed unknown before. The heckled ram, trying to shake off its tormentor, laid itself broadside open to the fire of the other Union boats. A shot crashed it, it was surrounded, forced to surrender. It was the cocky *Chickasaw,* unscathed, that led, on the following day, the peppering of Fort Gaines. All three forts were reduced within two weeks, the enemy fleet crippled, and the battle of Mobile Bay, one of the most brilliantly fought of the war, was over. The Gulf was cleared, the Mississippi safe.[22]

Stories of the ironclads' prowess had seeped to the remotest farm. Several other northern builders had furnished a river gunboat or so for the Union's mid-valley campaign, but it was of James Eads and his ironclads that men at village stores talked. Eads, the one-time diver, knew the Mississippi, top and bottom, he could walk to New Orleans on its bed if he wanted to. It took him to build the kind of boats that opened the river! In Washington the Navy heads were saying that the Eads gunboats, "the backbone of the river fleet," had gone far toward ending the war in the West, and that ocean fleets of the future might well be partly made up of armored monitors with the Eads type of turrets and fast guns.[23]

While the saga of his gunboats grew, James Eads, back home and in fair health, was far from enthusiastic about any of the ironclads he or anyone had constructed. Granting the good work they had done, they could have been much better. The design of armored gunboats should be greatly improved, he told his many callers. There could be far more effective warships than any that he or others had yet designed, he admitted to the Secretary of the Navy, reporting upon what he had learned in Europe. Better gunboats, he mused as he drove with his oldest stepdaughter, Genevieve, through the wintry streets of the Capital to have a Christmas dinner at the home of Attorney General Bates. Better gunboats, he pondered as the winds of March, 1866, scraped the tree branches against each other outside his library windows. He had taken out three more patents for gunboat equipment since the first of last year, but still more improvement was needed. Much more.

Then for a while Eads put gunboats aside. The refugees had crowded into his mind once more. The raw wind chilled even a well-clad man to the marrow, and the poor derelicts that had found their way here were in rags again, and underfed. The lately rich Sanitary Commission had been drained dry by its helpless human burden, and the present Commander of the Western Department—there had been a goodly succession of them—General Dodge, was about to revive the tax upon Southerners to raise an emergency fund of ten thousand dollars for "Asylum House," a refugee home. That hateful, mistaken tax brought to life again! James Eads went straight to General Dodge about it and begged to be allowed to try, himself, another plan before resorting to this one-sided assessment. He was confident, he said, that the banks of the city would much rather give the needed amount than to revive old animosities and thus defeat President Lincoln's plans for the pacification of Missouri. Dodge consented to wait. Within a few hours Eads wrote to the General that the officers of one bank, the Third National, of which he was a director, had heard his plea and confirmed the thousand dollars he had offered in their name. The other banks of the city fell quickly into line, the necessary fund was raised, and the unfair tax against a political minority was never revived.[24]

Hardly had Compton Hill greened another year than the tragic War Between the States ended, the last of the Confederate forces surrendering in May, 1865. In the following months James Eads was busy catching up with many things he had long wanted to do. His grounds had been neglected, his house needed additions and remodeling, new stables had to be built.[25] As one of the promoters of a vast City Park, he pondered notions for its development. He went to town on errands and brought home a carriage load of friends to dinner. He parried sly digs about a lively altercation he had had—and it had gone beyond hot words—with one William Renshaw, Jr., from whom he and William Nelson had bought some real estate some years ago. He defended Andrew Johnson who had succeeded slain President Lincoln. He smoothed over war rancors whenever he could. He talked of the importance of preserving the history of Missouri—it was notable and dramatic.

On August 11, 1866, James Eads met with a handful of others at the St. Louis County Courthouse to organize a Missouri Historical Society.[26] Presently he was deep in preparations for a visit of Andrew Johnson to St. Louis. The reception would be on a splendid scale, for it was the first time that a president had ever come here.

On the morning of September 9, a large reception committee and a fleet of thirty-six steamboats met the presidential party, which included Secretary of State William H. Seward, Secretary of the Navy Gideon Welles, Admiral Farragut, Generals Grant and Hancock, at Alton, Illinois, thirty miles up the river, and escorted the guests to the steamboat *Andrew Johnson,* which headed the fleet back to St. Louis. On the trip downstream, James Eads made the welcoming address, with a background chugging of many engines and splash of water over paddle wheels. His admiration for plain, honest Andrew Johnson rang in his words: "The citizens of St. Louis, irrespective of party . . . tender to you and to your illustrious counsellors and companions, the hospitalities of this city. . . . When you see their city, the vast creation of a few short years, reposing like a youthful giant by the side of its nurturing parent—the great Father of Waters— and witness the sinews of its strength and discover the secret of its growth, you will learn that its destinies are controlled by scores of thousands of earnest men, who are toiling with hand and brain in the upward and onward road of human progress. They have sent me with their welcome. . . .

"Your friends have witnessed with breathless anxiety your contests with the enemies of the Federal Constitution. . . . We have no sovereign in this nation but THE LAW—the written law; and he who is not loyal to that has treason in his heart. While all other officers of the Government promise to support that *aegis* of our liberties, you alone, sir . . . are required to swear that you will *defend* the Constitution of the Republic."[27]

The fleet arrived at St. Louis shortly after two o'clock and was received with wild excitement, cheering, cannonading, ringing of bells and screeching of whistles. The party was escorted in a long procession to the Lindell Hotel, where Mayor Thomas addressed it. In the evening there was a banquet at the Southern Hotel, "the *menu* for which filled a half column in the newspapers." Here President Johnson spoke at length.[28]

More and more, as valley commerce pushed rapidly forward, James Eads wanted to see the Mississippi navigation channels given some attention. What neglect the great river had suffered while it carried gunboats and troops, ordnance and food! It was time the government again took over the welfare of the stream that had served it so well. He talked a good deal about this, but the topic of gunboats kept shouldering it aside. He had built fourteen ironclads, seven musketproof tinclads and four heavy mortar boats for the Union, he had devised such startling innovations for them that half the world, it seemed, demanded to discuss the future of warships with him. He was still inventing better turrets and guns, arguing with John Ericsson over types, making models to show military men of his own and other countries. Baron Gerolt, of Prussia, was anxious to see the Eads turret adopted on the Rhine, the Belgian Minister wanted a conference with him on warships, Russian officials had the turret explained to them at dinners in New York and Washington.[29] Envoys from Washington or foreign capitals made their way across the country or ocean to sit before his fire and gather his ideas on the battleship of the future—and some of these ideas were so radical that only his unshakable belief in them made them plausible.

The day of the monitor was about over, he believed, especially for ocean use. Its low decks swamped too easily in a high sea. A gunboat of traditional vessel shape, its armor a very part of its strength and not carried as an extra burden, would doubtless dominate the waters of the world. But, unlike the old wooden broadside battleships, it would carry gun turrets of some type. As yet, no really fast, efficient, maneuverable seagoing warship had been designed, he admitted, and he did not expect to design it himself. Other men, salt-water men, would surely work it out —he was but a riverman who had pieced together, with unskilled workmen and haphazard materials, an emergency fleet for the valley streams. The improvement of the Mississippi was his first concern. His work lay there.

The river needed far more help than his Improvement Company had given it, there were levees to mend, shoals to reduce, besides removing litter from its bed. It was a matter for the federal government, he repeated to men who sent him their plaints about ruined channels or flood-drowned fields, and they would get government aid if they would pull together for it, instead of each section shutting out all but its own particular woes. There were boatmen who made it colorfully known that they did not give a "tinker's dam" if only the rooftops of the countryside showed above a flood, so there was a decent water depth in the navigation channels; and landsmen who could read with serenity about shoal-stranded vessels—a freshet would lift them out of the mud in good time. The Mississippi, James Eads argued, was not made up of sections, it was one stream. Moreover, it took adequate channels to bear crops and cattle to market, and it took land products to make up boat cargoes. The valley people should draw up a single over-all program and make a concerted appeal to Washington.

At last, early in 1867, the river convention that James Eads had so long desired was about to be held in St. Louis. And it happened that at this time he was informed by the Navy Department that he had again been appointed, with the authorization of Congress, to visit certain European countries, study their types of war vessels, report upon them and make recommendations. He consented to undertake it, but on one condition: that it wait until he had his say for the rivers.

The convention opened on February 12, 1867, in the Great Hall of the Mercantile Library. James Eads, looking over the assembly as he began his address of welcome, knew that he was at odds with most of the delegates. He wanted a valley-wide sweeping program, they wanted to undertake a little and hope for a little more. "The improvement of the Mississippi River and its great tributaries," he began weaving his accustomed multicolored gauze around a metal-hard intention, "involves the contemplation of one of the sublimest physical wonders of a beneficent Creator. This giant stream, with its head shrouded in arctic snows, embracing half a continent in the hundred thousand miles of its curious network, and coursing its majestic way to the Southern Gulf . . . has been given by its Immortal Architect, every inch of it, into the jealous keeping of this Republic.

"The garden which it beautifies and enriches contains 768,-000,000 acres of the finest land on the face of the globe, enough to make more than a hundred and fifty States as large as Massachusetts. If peopled as Belgium and the Netherlands are . . . it would contain . . . nearly one third of the population of the world."

Care of these streams that were so vital to the country in both peace and war should be demanded, he said. There was not a single river problem that engineering skill could not master. Then, his words coming like hammer blows: "Not a dollar should be voted by the representatives of this Valley for any public works while these great rivers remain neglected."[30]

The sharp vehemence of his speech having sounded a keynote for the middle country, James Eads was content to leave for Europe. As the inventor of rotating turrets and steam-controlled guns that spit shells with the patter of hailstones, he was again given warm social and official welcome wherever he went. And he became at once the center of diverse opinions. Adroitly he turned the opponents upon each other—he had not come to argue in favor of any one type of gunboat or turret, but to see and hear. There was much to hear, for the controversy over the merits or shortcomings of the turreted monitor and the broadside-firing vessel was loud and voluble. Stubborn prejudice buried all sound conclusion. Only in England did Eads find a calm view, and that was firmly in favor of the traditional broadside-firing warship.

He came home convinced that an entirely new model of warship, combining the best features of the two extreme types, should be developed. It would take him some time to work out designs for it and get up his report, for it would have to be done in and out of two consuming interests: a new project and the marriage of his older daughter, Eliza Ann, now twenty-one, to young Major James F. How, son of his long-time friend, ex-Mayor John How. "Blue-eyed Eliza," one of Martha's poems had called her —she was only six years old when Martha was taken from him. Her wedding was a great occasion to him, anything was that concerned his daughters. He would spare no trouble for it.

He busied himself over an invitation list that grew and grew. He could not leave out any of his friends, and there was such a host of them. This posed another problem: although his house had been greatly enlarged when he had had it recently transformed to an Italian villa by a local architect, George I. Barnett, it would not hold this throng. With something of the speed and efficiency that had provided a gunboat fleet for the rivers, he had a temporary ballroom put up beyond his banquet hall, handsomely decorated and chandeliered.[31]

When the wedding day, December 3, 1867, arrived, he had an awning stretched from Trinity Episcopal Church to the curb, and carpets spread under it to protect trailing dresses and slippers. The weather was vile, it rained incessantly. But so thoroughly had Captain Eads caught the public imagination that the marriage of his daughter was a public event, and crowds stood in a downpour to watch carriages roll up and the occupants alight. The street was blockaded with vehicles, "and the clattering hoofs of the horses sounded like the approach of a cavalry regiment." It took a police squad to hold back the curious and direct the tangled traffic.

After the ceremony the carriages filled one by one and rolled out to Compton Hill. James Eads, in careful evening dress, stood beside Eunice, who was always imposing in her blond, stately beauty, at the end of the long drawing room and presented the guests, eight hundred of them, who filed past to felicitate the petite bride in her voluminous white satin gown and the soldierly groom, elegant with his white-satin-lined low-cut vest and white satin tie. An orchestra was luring the overflow of guests into the dance hall, while the Southern Hotel chef and his staff prepared a midnight supper. When he could get away, Eads took some of his friends up to the large room over the "reception parlor" where the bridal presents were on display. There among gold-lined silver tankards and goblets and costly works of art was a small, modest gift—a wineglass made in his ill-fated factory. Martha had given it to a friend with the request that it be presented to Eliza Ann or Mattie at marriage. It was a long, gorgeous evening. Eads had never been more gracious, or his heart more full. Almost too soon the music and mirth died away and the carriages that had glutted the driveway and street were gone.[32]

In a few days the last guest from a distance had left. The house seemed empty without Eliza. And Genevieve was gone, too. She had married John Ubsdell, who was of New York and Southampton, England, but who had interests in St. Louis and a consuming fascination with the jetties and Port Eads. She was kept much on the wing. James Eads settled again to writing, at every possible free moment, his report to the Navy Department on European gunboats and his recommendations for American war vessels.

Approaching the subject from its simplest fundamentals, he began his argument: "The sole purpose for which the ship is constructed is to carry the battery. How important, then, that this battery . . . should be the most powerful in character that the ship can sustain. . . . The most powerful battery will be that which can throw the heaviest metal with the greatest force and rapidity, and with the surest aim."[33]

The warship of the future would, he thought, be shaped like the familiar broadside vessel, but would be provided with immovable (not rotating) turrets within which the guns, manipulated by compressed air or steam, could command the entire horizon through very small portholes; the rotating turret, he had become convinced, was too vulnerable where it sat into the deck. All the advantages of both of the disputed types of warships could be embodied in the vessel.

"I would advance this assertion with great hesitancy," he continued his report, "were I not able to submit to the department the accompanying drawings, showing at least one method by which they may be attained. . . . Although they are the result of much study on my part, they are not submitted as the best system that can be devised, but mainly in the hope that they may suggest to abler engineers and constructors that perfect plan we all desire."

In his sketches of a compound revolving battery within a fixed turret, he showed that the saving of armor area would permit much thicker metal plates. It was his belief, he said, that the design could eventually be developed so that armor twenty-four inches thick would be practicable. Inevitably, as time went on, every device to thicken the armor would be tried. Then, moving on to a sensational prophecy, he concluded:

"It is quite possible that the limit to which this will be carried in the future will be reached by submerging the entire vessel at will, leaving only the turrets above water in action. By such a system hull armor will be unnecessary. The engineer who is willing to ridicule and deny the possibility of our posterity overcoming the difficulties and dangers that now prevent submarine navigation from being made entirely safe for commerce and eminently practicable for purposes of war, has less hope and confidence in the scientific and inventive genius of man than I have."[34]

It could hardly be denied by anyone familiar with the later fixed-turreted ocean battleships and their rapidly maneuvered guns that they bore the imprint and rich inventiveness of a Mississippi riverman.

CHAPTER VII

SPANNING THE BIG RIVER

ALTHOUGH, at the close of the war, James Eads had entered heartily into devising future types of ironclads and naval guns, he was resting, in his way. His repose appears never to have been inactivity, but, rather, the browsing of his mind in whatever pastures lured it. He managed to live in the two worlds of actuality and creative abstraction without the conflict suffered by most men of genius, he could work and dream with a constant flow of small everyday human affairs around him. The comings and goings of his daughters, their lively chatter, the rustle of crinoline, a grind of carriage wheels, the tinkle of a pianoforte, were a mosaic background for his prodigious concentration. In the evenings men of widely different interests gathered about him to talk of many things: of the gun-turret models he was putting together, of the rapid recovery of steamboating since the war ended, the growth of valley agriculture, the colorful settlement of the Northwest, the chimera of a St. Louis bridge across the Mississippi, replacing the ferries.

The most matter-of-fact items would shift vivid pictures across Eads's graphic fancy: the new majestic steamboats that were coming out on the rivers, their paddle wheels as high as a waterfront warehouse, their chimneys rearing far above; the prairie wheat rippling to the horizons, the reapers made by Cyrus McCormick, a one-time Virginia blacksmith, clacking through it; gold seekers crowding a boat for Fort Benton, two thousand miles up the shallow Missouri River and a half mile above the sea, a lusty village where miners, missionaries and desperadoes jostled each other in tough saloons and tawdry dance halls;[1] a bridge across the Mississippi—it was high time for it.

A small bridge company had organized a year and a half ago —it was now late summer of 1866—and secured a charter from the state of Missouri. But when they turned to the Illinois legislature for a like permit, the Wiggins Ferry Company, of St. Louis, which enjoyed a near-monopoly of carrying freight and passengers across the river, had waged a fight upon the bill, getting it loaded with such cramping amendments that the finally granted charter appeared almost worthless. Still some of the company's officials had taken it to Washington for the necessary Congressional consent to span a navigable stream, and there, no more than a month ago, in July, the federal approval had been given. Little good it would do, for, in spite of all that Gratz Brown, now in the Senate, could do, more restrictions had been added.[2] It was impossible, everyone said, to bridge the Mississippi under such conditions. "Impossible, impossible" had echoed ever since.

The word challenged James Eads. Why should a bridge be impossible? Progress always found a way, he held stoutly. Take the steam ferries, for instance—a prophecy of them would have sounded mad seventy years ago when enterprising Captain Pigott planked two pirogues together and had his French boatmen pole or paddle them across the stream, their lithe bodies and monotonous chant marking the rhythm. Ferriage was expensive and troublesome, at best, with freight having to be dumped from train or wagon on one bank and loaded into train or wagon on the other—then there were times when the river was low, or stormy, and the ferries could run safely only from dawn to dark. And there was midwinter when ice locked out service for weeks while freight piled up at the wharves and distracted travelers fumed. With railroads multiplying on each side of the river, there would have to be a bridge they could cross.

Rivermen, and there were always some at hand, flared up at this. It was steamboats that had made St. Louis the metropolis of the valley, destined to become the nation's first city, perhaps the capital.[3] Its splendid future was foretold in every sign. Boats heavy with gold followed each other here from the mountains, lumber raft "cribs" from the north were often as thick in these waters as Indian war canoes, cattle were driven from as far away as Texas to this central market, and the autumn hog drives verging toward it congested the country roads. This was the most sophisticated city in the land. Where else was there such a rampant Woman Suffrage group? Or a comparable Philosophical Society —its teachings were coming to be known afar as "The St. Louis Movement," and even Ralph Waldo Emerson and Bronson Alcott were turning to this fountainhead of truth.[4] All that without a bridge!

Riverman though he was himself, James Eads contended emphatically for a bridge. The great stream that had so blessed commerce moving north and south should not remain a barrier between East and West. Moreover, if a bridge was not put up, railroads would branch around through Chicago and over the one or two bridges that had been thrown across the shallow upper river. As for himself, his prime interest lay in making the Mississippi a safe, deep waterway to the sea.

He was likely amused at the explosive effect the mention of Chicago had upon his friends. Chicago, the metropolis of Garlic Creek! Its blowhard drummers overran the country selling the most outrageous catchpennies, even knockdown houses and churches.[5] It was Chicago railroad lobbies in Washington that had done most to burden the St. Louis bridge charters with new musts and don'ts, so that now, added to the Missouri legislature's ban on suspension and draw-span bridges, and Illinois amendments, there was an order that at least two of the spans must be 350 feet wide or one of them 500 feet wide. The first would make too much expensive masonry work in a rushing stream, a 500-foot span was impossible.[6] Railroads . . . steamboats . . . Chicago . . . a bridge, a bridge, rose in clashing refrains. Faces were flushed, chair arms pounded.

Into this discord the host would toss an absurd riddle or a timely yarn. "No one knew more droll stories," a grandson of James Eads relates of him. He had a fresh one for nearly every occasion and told it with gusto, laughing at it himself "in the most infectious way till he was red in the face. Indeed, he was the larger half of his stories."[7] He followed tale with quip until the room rang with merriment, all differences put aside.

The matter of a bridge engaged Eads more and more. He had taken it for granted that a bridge would be built presently. Men had been talking about it for years. In 1839, after the mile-wide river had been narrowed by connecting Bloody Island with dikes to the Illinois shore, Charles Ellet had proposed a suspension bridge for vehicles and cattle—there was no thought at that time of a railroad's ever reaching here. When the eastern "railroad mania" began creeping westward a rash St. Louisan, Josiah Dent, had startled the community, in 1855, by prophesying that railroads would soon stretch to the Mississippi, and want *to cross it*. He had excited a few friends into a bridge company and got an engineer to plan a structure, but the project died of fright.[8] In a twinkling of time the Baltimore & Ohio Railroad, by means of short links, did reach the Mississippi opposite St. Louis, in June, 1856, at Illinois Town, a drab ferry landing flanked by a few warehouses and cottages. Restaurants and saloons had sprung up there, plank walks were laid in the rich black mud, and the town was renamed East St. Louis.[9] Other railroads had come to chin the east bank, demands for a bridge rose on all sides, but the national tumult over secession, and then the war, had pushed them aside. Now the clamor was swelling again. A bridge, a bridge!

A bridge. James Eads dabbled at sketches, making his unorthodox rule-of-thumb calculations. An arch bridge with spans of five hundred feet would meet the legal requirements. Impossible? Men often tossed that word about when they meant "unprecedented." There had never been a bridge arch of more than half that length,[10] but that fact was beside the point. Three long arches, with short arch approaches—such a structure would be a thing of impressive beauty, worthy of the continent's greatest river.

Talk of a bridge, most of it gloomy, rose in fitful gusts.

Financiers and constructors would not touch the St. Louis project, bedeviled as it was by engineering hazards and cramping laws. They were especially daunted by a clause in the Illinois charter demanding that the east approach be through a congested business section of East St. Louis where a right-of-way would be ruinously expensive. And there would be costly rights-of-way in St. Louis. Eads bent his full attention upon his drawings and figures. Enough time had slipped by in futile talk, the bridge must be built.

"In taking over any project," another engineer wrote of him afterwards, "he gave it long, careful and thorough examination, looking at it from all sides. . . . When once his mind was made up, it never changed; once having stepped forward he never took a backward step, no matter what obstacles confronted him; his faith never wavered. . . . He never became discouraged for a moment, however dark it seemed."[11] He stepped forward now to press the building of a bridge. The little company was heartened —Captain Eads had cleared the river of wrecks and driftwood, he had provided it with a navy, and he would get someone to bridge it.

At this time a bright light of hope dawned upon the company's horizon in the person of Mr. L. B. Boomer, a bridge contractor from, of all places, Chicago, the braggart town whose railroads "pierced every portion of the northwest," and that lately had shown the effrontery of claiming a population of two hundred thousand, within four thousand of St. Louis! Mr. Boomer proposed not only to design and build the bridge, but to get the Illinois legislature to strip the charter of its villainous amendments.

The wondering but grateful company agreed to all this, and sent a committee to Springfield with Mr. Boomer to put through a bill dooming the amendments. Their bill passed readily in the lower house. Oddly, Boomer seemed surprised at this, and thoughtful. He urged the committee to return home and leave the rest to him. They did, and waited to hear that the bill, oiled by the influence of their new aide, had slid through the state senate. The news that came stunned them: Boomer had side-tracked their bill and substituted two others, one to repeal their charter outright, another to give a company of Chicagoans an *exclusive right* for twenty-five years to bridge the Mississippi at St. Louis.[12]

The news licked through the streets like a prairie fire. Aghast bridge enthusiasts rolled out to Compton Hill with it. St. Louis had been betrayed, taken in by a slicker from "voracious, overgrown" Chicago. It was clear enough now that Boomer had never meant to build a bridge, but only to make sure that none would be put up at St. Louis for a generation, at least. James Eads was disturbed. The charter, such as it was, must be saved. A committee of three was dispatched to Springfield to do it. While this delegation worked like beavers, name-calling in St. Louis reached one of the highest marks since the early French settlers wearied of villifying the Americans who had bought from Napoleon Bonaparte the very ground under their feet. Eads worked absorbedly over his sketches.

The charter was saved, but Boomer's *exclusive right* bill threatened to pass. James Eads set his drawings aside to attend an emergency meeting of the Bridge Company, where he acted as chairman. He heard the indignant accusations, noted the hot anxious faces, and dispatched the delegation back to Springfield to demand of the legislature what right they had to consider such a bill in the face of their own still valid charter. He could chuckle later over the dismayed explanation of some of the lawmakers that they had thought the two bridge companies were virtually one, cronies, fond and inseparable. He grew speculative when outcries reached him that Boomer's *exclusive right* bill had crept through to victory. On a march evening, 1867, he drove into town to another desperate conclave of the company's few stockholders, at the Planters' House, carrying a large rolled paper.

He stood before the frustrated men, his slender body tense, and told them in a few vigorous sentences that they would have their bridge. Every obstacle in the way of it could be overcome. Their charter was still in force and Boomer's *exclusive right,* therefore, worthless. All legal restraints and demands could be met in the design: the terrifying cost of rights-of-way could be cut down by elevating the tracks over the business section of East St. Louis, and by tunneling just under the high Third Street level in St. Louis—the vehicular road of the bridge would be set above the railroad tracks exactly meeting the street level. Commerce from ocean to ocean would feel the benefit of a bridge across the Mississippi here. And the bridge would be built.

Borne along with the intense conviction that surged in his voice, his hearers watched the roll of paper unfurled and hung on the wall. The Bridge! They formed a crescent about it, tiptoeing, peering. It was impressive, four massive foundations supported the tips of three lacy arches curving over incredibly long spans. Grace of line and proportion were emphasized throughout. The two middle piers, Eads explained, would be 520 feet apart, the other two piers, or abutments, would stand 502 feet shoreward from these, and at high water would be well out from shore. Three of these foundations could be sunk to bedrock, the fourth one, near the east bank where the stone floor lay extremely deep, would be set on metal piles driven down to it. Arches of this length, although never existing before, were entirely practicable. These would be of steel.[13]

Piers carried to bedrock in the tumultuous river! Arches of steel—steel! Who had ever heard of a steel bridge? Their Bridge would be magnificent, its building epochal, the eyes of the whole country, of the world, would turn upon their city. Buoyed by the first optimism they had ever enjoyed, and gloating over the certain discomfiture of Chicago, the company adopted the design and appointed James Eads as chief engineer.

Even the news that Boomer's *exclusive right* bill had been passed and signed by the Illinois Governor did not riffle the complacency of the Bridge promoters. They urged Chief Engineer Eads to proceed at once with his preparations, and cast about for a consulting engineer to whom he might turn for advice. Mr. J. H. Linville, president of the Keystone Bridge Company, of Pittsburgh, who had contracted to erect an iron bridge across the Mississippi at Keokuk, was selected, and the Eads skeleton sketch sent to him for his possible suggestions.

The sketch came back at once with Mr. Linville's blunt statement that he would have nothing to do with such a design, a bridge built according to it could not support its own weight: "I cannot consent to imperil my reputation by appearing to encourage or approve its adoption. I deem it entirely unsafe and impracticable," he wrote, and enclosed a hastily drawn substitute plan. The Bridge Company officials were affronted by his offhand condemnation of their design, and at what they considered his presumption in offering them a makeshift, ordinary structure. They might have been shaken by his emphatic disapproval, except that James Eads stood so firmly by his design. Linville and his threatened reputation were set adrift and no consultant was hired.[14]

Weeks passed in which the Chief Engineer all but vanished from sight. He was submerged in working out intricate details for the Bridge, solving each mathematical problem in his own fundamental way, without benefit of formulae. When he later had two engineers, Colonel Henry Flad and Charles Pfeiffer, whom he had chosen to assist him, make the computations by academic rules they arrived by their methods at about the same conclusions that he had reached by his. In the middle of July, 1867, he handed the Bridge Company's board of directors a full outline of his plan. It was immediately printed and published afar, setting railroad men all over the country to speculating about the effect of this crossing of the Mississippi, and engineers to shaking their heads over the design: foundations sunk to bedrock in a raging river, arches five hundred feet long, and of steel. A Frenchman, Jules Verne, had just written *Voyage to the Center of the Earth,* and *A Trip to the Moon*—fantasy was all the rage. James Eads must be a man of colossal imagination.

While Eads was devising the equipment for sinking the first bridge foundation and having it constructed, Mr. Boomer loomed again. He had formed a bridge company in Missouri, himself owning three-fourths of the stock, and had this firm assign all its holdings to his Illinois and St. Louis Bridge Company—a name easy to confuse with that of the St. Louis and Illinois Bridge Company of Eads and his friends. Boomer had paid himself ten thousand dollars of his company's money for the transaction, elected himself president and treasurer, and was now spreading a statement to the four winds that only his company could tie a bridge to the Illinois shore opposite St. Louis. A bridge put up by any other group would have to be left dangling, unfinished, over the stream.[15]

Wrath rumbled along the west bank: If Boomer built a bridge it would be controlled by Chicago. That desperate city, bent upon consoling itself for having just failed in an attempt to deepen its canal enough for large ships, would stop at nothing. "Chicago, that Babylon of houses that fall down, reaching after trade to support its fast horses, faster men and fallen women. Beware, O Chicageese!" warned the St. Louis *Missouri Democrat,* of May 1, 1867.[16] Suspicion that its rival was trying to wax and fatten by blocking the construction of the Bridge was heightened by a near-gale of enthusiasm in the Chicago papers over Boomer's company—anything fostered by that press was sure to be of baleful intent.

All this surged back and forth past the Chief Engineer, but he was too preoccupied to give it much heed. In August he gathered pile drivers, workboats, derricks, engines and barges of material in the river off of the foot of Washington Avenue, midway of the long St. Louis waterfront, where the West Abutment, or pier nearest the west bank, was to be put down. Boomer took disgruntled stock of this, the persistence of that fellow, Eads, was galling. A blunter attack upon him would have to be made, a public discrediting of his plans, a drying up of any financial support for which he might hope. Hurriedly Mr. Boomer called a convention of twenty-seven engineers, many of them his own employees, to open in St. Louis on August 21, 1867.

It had pleased James Eads to begin, on August 20, as the convention delegates were arriving, the construction of a cofferdam at the West Abutment site. Now billows of smoke spouted from engines into the sunlight, their black shadows curling over the chopping shimmer of water. Cables creaked, mechanical "travelers" carried their burdens to place, and the first mighty blows of a pile driver thudded out, drowning the ferry whistles, terrifying carriage horses and dray mules, belittling the rap of the convention gavel in the parlor of the Southern Hotel, and punctuating the speeches ominously.

But Mr. Boomer's engineers proceeded with splendid concentration, for, only five days later, the *Missouri Republican* disclosed that their investigation had "covered the entire ground; the most intricate problems had been patiently solved; the strength of materials estimated; the safe length to which spans might be extended calculated; the comparative value of different kinds of piers and foundations determined upon; and every point connected with practical bridge-building elucidated in the light of the most thorough knowledge and the most varied and extreme experience." What more could any convention of experts have accomplished in any stretch of time?

In close harmony the engineers had chanted that any attempt to set piers on the bedrock of the Mississippi at St. Louis would be perilous and futile. They chimed a resolution that, while their decisions did not have the least "reference to any particular company," they harbored "unqualified disapproval of spans of five hundred feet . . . for which there is no precedent." Mr. Boomer had displayed a twelve-foot drawing of a truss bridge of six spans, and the delegates had heartily endorsed it. His convention was a rousing success and well worth the sixty thousand dollars it had cost his backers. Its report was printed in an elaborate pamphlet and scattered over the country. The effect upon capital was profound: a body of engineers had virtually condemned the unusual bridge that James Eads was commencing to erect at St. Louis! The impact of this report delayed the completion of the Bridge more than a year, it was later estimated, and weighed its builders under a heavy burden of interest.

In reply to the convention's criticism of his plans, Eads toweringly demanded: "Must we admit that because a thing never has been done, it never can be?"[17]

With little waste of argument, he exhibited his Bridge sketch at the Merchants' Exchange where it could be compared with the plan endorsed by the convention. The sketch won its case. Those who came to carp at it admitted that the arch Bridge, with its wide spans and graceful lines, would be an inspiring structure—if it actually could be built. Even steamboatmen, never susceptible to bridges, however charming, expressed themselves as pleased that this one offered the least possible drawback to navigation. But elsewhere in the country suspicions of the design chilled interest in the project and left capital congealed. Still James Eads went stubbornly on with the construction, his associates following him in implicit faith.

Mr. Boomer scowled over this phenomenon. He was convinced that nothing but antagonism to him as a Chicagoan would keep these men so doggedly at an unfinanced venture. He would have to disarm them. To this end he resigned his offices in his bridge company, filled them with St. Louis men willing to go along with him, and from his now blameless position wrote to a St. Louis newspaper that he certainly did not represent Chicago in any way. But he could not resist topping off with a blistering condemnation of the Eads plan, adding that the St. Louis company using it had no legal existence in Illinois and that its pretensions might well be hauled through the courts of that state.[18]

James Eads was beset by his associates to answer this. Replying bluntly to the effect that his critic knew next to nothing about the design he so glibly condemned, he went on with his work. His friends, however, took Boomer's latest outburst much to heart. They had not been mollified by his placing St. Louisans in his company, a move they suspected as a sinister ruse—only the worst motives were ever ascribed to Mr. Boomer. His threat of court action was discomfiting, it was hard enough at best for them to keep their slight toe hold in the confidence of the public and press.

At this moment there was fairly thrust into Boomer's ready hand a weapon impossible to dodge: James Eads had run into serious difficulties in his work at the West Abutment.

CHAPTER VIII

BELOW THE RIVER BOTTOM

The opaque waters that rippled over the spot which James Eads had chosen for the West Abutment hid many a secret. For sixty years they had swallowed every kind of discard from boats: lengths of smokestack, worn-out grate bars, sheet-iron furnace envelopes, fire brick, and various other things. Two of the twenty-nine boats sunk in the great fire of 1849 lay submerged here, one hull partly covering the other. Moreover, the city had widened its wharves after the fire, building solidly over the end of the two wrecks, and while this extension was in progress another vessel had sunk here. Nevertheless, Eads was positive that it would be best to stretch the Bridge from this point near the middle wharves so that boats from the north could stop above the structure and those from the south could stop below it, for even two piers in the river channel would offer some obstruction to navigation.

The site was to be enclosed in a cofferdam formed of two courses of sheet piling driven down through the sand bed of the stream to bedrock, which at this side of the river was only 47 feet below a level accepted as "high water," the height of a memorable flood in 1828. This level, marked by a curbstone, known as the city *directrix*, at the foot of Market Street, served as a datum plane for all city surveying and for nearly all references to river heights or depths. There was also a plane, "extreme high water," the level of the greatest of all floods, that of 1844, which was sometimes used by St. Louisans in superlative description.[1] One course of the sheet piling would sit inside of the other with a 6-foot margin between, like a box placed loosely within a larger box, the margin to be filled with clay. The water would be pumped out of this strong enclosure, all debris removed and the sand excavated.

To cut through the wrecks, James Eads had a monster chisel secured in the end of a heavy oak timber, and into the cleft that it made in the wrecks the piling was driven. When he had completed the courses, had the space between them clay-filled and the water pumped out, he discovered that some of the piling had failed to pierce entirely through the lower wreck. He set men to patching the gap between this wreck and bedrock, but before they could complete this, water had broken into the cofferdam, flooding it. A new course of piling was driven down inside the dam, and a diver sent down to remove, piece by piece, whatever parts of the vessels remained within the enclosure. After this margin was filled with clay, the water pumped out again and the excavation got under way, several steam engines and the wrecks of four large barges were found within the dam.

It was a bad start and furnished much adverse publicity. Mr. Boomer and his aides kept the world apprised of each new difficulty at the West Abutment. "The dam leaks again, water boils up in it faster than two pumps can remove it," they cheered in print. "The Eads Bridge will cost three times as much to build as ours, and will take three times as long."[2] The project was now considered a harebrained scheme, wags grew facetious over it. An item in the St. Louis *Missouri Republican* ran:

A gentleman crossing the river by ferry pointed out to a stranger the site of the prospective Bridge. "How much will it cost to build the Bridge?" the stranger asked.

"Seven million dollars," the other exaggerated.

"How long will it take to build it?"

"Seven million years."[3]

The doubts that glared at James Eads from newspapers were prickling, the suggestion by well-meaning acquaintances that he abandon the hopeless venture was galling. Gravely he was warned that his attempt to set foundations on bedrock would end in disaster, and that steel arches a tenth of a mile long, with only their tips resting on masonry, would collapse of their own weight, even if the piers did not crumble under them. But hardest to bear, doubtless, was the sympathy looking from the eyes of loyal friends, the silent verdict that his whole plan of the Bridge was the pardonable fancy of an imaginative, generous man.

The Bridge was imperiled by universal doubt. At any time, Eads feared, the small band of promoters might succumb to the bombardment of pessimism. He would have to convince them that his vision was more than a zealot's vagary, that his plan was sound even though the world had never known just such a structure as he intended building. He decided to write to the people of the nation a full description of the Bridge, addressing the script to the Bridge Company as a report. "Anyone who can be made to understand the principles of the simplest of all mechanical powers, the *lever,* can readily comprehend the explanation I propose making," he began this plea.[4]

He looked up from it when Mr. Boomer, having found that words pattered off of his obstinate rival, tried a flourish of action, signing some contracts (reserving the right to suspend them at any minute), and starting an excavation in the East St. Louis levee. It was too bad that Boomer had neglected to acquire a right to the site, for his laborers were arrested for trespassing. After he had them freed and got them back to work, they were arrested again, and again released. Withal, however, quite a hole was dug in the east bank, and eighty timbers were dumped beside it. Annoyed that Eads appeared to take little notice of this, Boomer shifted his strategy, threatening to have the officials of the St. Louis company hailed as culprits into an Illinois court for assuming that they had a right to base one end of their Bridge on the east shore—what with retrials and appeals, the Eads project could be held up for years.

At once St. Louis quit scoffing at the Bridge and rallied to its defense, accusing the Chicagoan of everything reprehensible, including blackmail. Boomer was taken aback by this flare-up and wrote a hurt letter to the St. Louis *Missouri Democrat* denying that he was a blackmailer, he was seriously building a bridge with "a large force of men employed, not amusing themselves for months past in an idle effort to pump out the Mississippi."[5]

At this moment James Eads found it convenient, and likely diverting, to tread on Mr. Boomer's coattails by having a bit of work done on the Illinois bank. At once all the officials of the Illinois company were ordered into an Illinois petty court on a writ of *quo warranto*. This action fizzled out and Eads's men kept on at their work. In the cofferdams across the river picks and shovels were nearing bedrock. On February 25, 1868, the St. Louis company held a ceremonious cornerstone-laying, the Chief Engineer adjusting to place the massive limestone block that was lowered to the bottom while spectators on boats and ashore applauded.[6]

All this was encouraging, but James Eads knew that the project was in danger. Beneath the surface of the rival companies' petty backbiting seethed an earnest conflict that might move on to a fatal deadlock unless some peaceable agreement could be reached. He arranged, through Congressman William A. Pile, to meet the president of Mr. Boomer's company, David Garrison, in Washington, far away from local strife. There the two men, both St. Louisans, both keen for a bridge, whittled their antagonisms down to a single difference: the type of bridge to be built. They agreed to submit their respective designs to an impartial board of engineers, and that the company whose plans were found the less desirable would sell its franchises to the other and retire from the field.

Weeks later, when the smoke of a considerable battle had blown aside, a single company, made up of the two groups merged, stood forth. It bore the name of Boomer's company, but it flaunted the Eads arch bridge design as its chosen plan and had James Eads himself for its chief engineer. Enmeshed and miserable, Boomer and his friends one by one withdrew from it.

All that was necessary now in order that the Bridge construction could go sailing along was for Congress to approve the merger. James Eads set confidently about getting this done, but found lined up against him every kind of opponent of any kind of bridge at St. Louis: the northern railroads, rival river ports, and odds and ends of legislators who were temperamentally against anything brought up in Congress. These varied opponents got together and agreed that the vulnerable point in the St. Louis project was the Eads design. Engineers had declared for it, but financiers could be easily frightened about it. They concentrated their attack on it.

While this assault was gathering force, James Eads rushed the printing of his Bridge report, publishing it in May, 1868. Simple, but profound, it was the story of a dreamer's steel-strong intention. Bridging this boisterous stretch of the Mississippi had been dallied over long enough—he would bridge it. He had delved into existing theories of stress and strain, materials and methods, and far deeper for new concepts. So bold and original was the report that it was "eagerly read by engineers and financiers the world over," and scientific journals lauded it. Eads felt certain that it would win for his Bridge the skirmish in Congress, but the wave of interest in this unusual document veered far around the legislators who were sparring over his bill. It was an unflattering opinion of him and his plans that finally carried the merger bill to victory, some of his opponents having decided that the madly impracticable design would itself bar completion of the structure—no arches of five hundred feet ever had or ever would be built, and the attempt at it would keep James Eads and his henchmen busy and out of mischief for a long time. They leaned back and let the bill squeeze through.[7]

The crisis was over, yet disappointment tugged at James Eads as he sat before his desk in the library at Compton Hill, looking away past the wooded slope to where the river lay against the far horizon. The Bridge would span the Mississippi, but he would not stretch it there himself. His vigorous defense of it in Washington had drained his strength, and there had been his anxiety over his father. The handsome old colonel, with his shock of white hair and thick short beard, his distinguished manners and lavish spending, had reached the end of his colorful way. A conspicuous figure in Davenport, where he had moved after Nancy died, he had married twice again, most unhappily—all women were not as patient as Nancy—he had tangled his financial affairs as usual, and fallen ill. James had brought him to St. Louis, cared for him through his remaining months and laid him to rest in Bellefontaine Cemetery, overlooking the river.[8] Shaken by a ceaseless cough, the Bridge's Chief Engineer fenced with fatigue and lost. Always, it seemed, when his whole being cried out to build, build, his body gave way. Now the doctors were bundling him off to sea again, where he could find no work to do. He had to grip himself to make the words of a resignation from his Bridge Company post sound as though they were not dipped in the bitter brine of frustration. He suggested W. Milnor Roberts, a well-known Pennsylvania engineer, to fill his place.

The immediate reply of the Bridge Company filled him with warmth. They would have none of his resignation. They had undertaken a stupendous work, and only his fighting courage could put it through. He would be given a leave of absence, with Milnor Roberts substituting for him until he came back.

It was a depressed group that James Eads left behind as he crossed on the ferry to take a train at East St. Louis. The Assistant Engineer, Colonel Henry Flad, complained that with the departure of his chief "all the life of the company seemed to go out." It was only a short while before they were making ready for his return, for he had landed in New York in December, 1868, and would pause there only long enough to combat some lingering distrust of his plan among financiers. But so headlong did he fling himself into this effort that he was hustled back to sea on the stern order of his physicians—not, however, until he had raised a fat sum in stock subscriptions for the bridge. Again the little company drooped and waited.[9]

In Europe their chief engineer rested in his usual way, staving pell-mell into the invention of a sand pump, and journeying here and there to confer with bridge experts and to look over the metal market. In Paris he met M. Moreaux, known as "the builder of a thousand bridges," and actually the builder of ninety.

M. Moreaux generously consented to glance over Eads's sketches. He found the design so excellent that it "could not be improved," yet there was one cardinal fault in the whole matter: the astounding presumption of an American engineer in publishing plans for a bridge having spans more than two hundred feet wider than any arch bridge in Europe! Such a bridge had never been attempted before, even by himself! At such a moment a smile might narrow Eads's eyes, but the wry remarks that leaped to his tongue were held back by his straight lips. His caustic humor was seldom loosed except in public debates. M. Moreaux, having eased his professional indignation, invited the American to visit the site of a bridge he was building at Vichy, where piers were being sunk in the River Allier. They set out together, two daring engineers engrossed in bold undertakings.[10]

This jaunt was the high point of James Eads's busy rest in Europe. The piers in the Allier were not being sunk in cofferdams, but by the *plenum pneumatic* method, the excavation being done in huge diving bells, or caissons. Eads was enthusiastic about it. By this system the very deep East Abutment of his Bridge might be based on bedrock, the sinking of the two channel piers would be greatly simplified. But the Mississippi was not the mild Allier, and he would have to initiate drastic changes in method and equipment.

Stopping in England again on his way home, James Eads explained his Bridge plans to engineers there, so impressing them that they proposed him as a member of the Royal Society, the oldest scientific association in Great Britain.[11] He returned to America in April of 1869, much improved in health, and soon appeared before the stockholders of the Bridge Company, buoyant as he told of the compressed-air method by which he would sink the deep channel piers, and possibly the East Abutment. Although it had never been used in America, this plenum pneumatic system was not new. In fact, the idea of carrying on underwater construction had been conceived by a French physicist, Denis Papin, in 1674. About a hundred years later, in 1779, Charles Augustin de Coulomb presented a plan by which work could be done thirty or forty feet underwater by this method. Lord Cochrane, of England, had patented a compressed-air device in 1831, and William Bush had invented a caisson with an air chamber in 1841. Later, Fleur de Saint-Denis had used compressed-air in the construction of a bridge at Kehl, over the Rhine. M. Moreaux was employing it now on the Allier.[12]

While the system was old and tried, Eads said, it would have to be used in the building of the St. Louis Bridge on a scale never attempted before, as no other underwater foundations in the world had ever been so massive or so deep. Borings showed that the East Pier, in the channel, would go down about 120 feet below high-water mark—equal to the height of a ten-story building—to reach bedrock, and the East Abutment would be based much deeper. Because of these factors, and of the violent vagaries of the Mississippi, distinctly new types of caissons, floating apparatus and machinery must be invented and put together.[13]

The brief report was breath-taking to the listeners. They were awed by the tremendous nature of their project as the resolute voice of their Chief Engineer brought it home to them—masonry of stones, weighing tons each, being set up about the outside of the caisson as it gradually sank through water and sand for more than a hundred feet! Only the inspired confidence of the slight man with the deep, lighted eyes made the task seem feasible. When James Eads wrote, he was an engineer and wit. When he lectured, he was an engineer and crusader, charged with emotion, sweeping his hearers along with him.

From that day all was bustle and activity. Eads, at his town office, wrote, drew, ordered, hired, dispatched messengers. He was having limestone quarried at Grafton, Illinois, upriver from St. Louis, brought downstream and stored on the levee; granite dressed at Richmond, Virginia, to be sent via New Orleans; the caissons built by William Nelson—his former salvage partner and aide in gunboat construction—at Carondelet; large barges made ready to be fitted up with derricks and travelers, air and water pumps and various engines; his sand pumps made and tested, brick and sand heaped on the waterfront, a blacksmith boat set up, timbers for piles, scaffolds and caisson girders dressed; tugboats put in order, and the breakwaters above the foundation sites constructed.[14]

These breakwaters, the first essential, were crude forts in midstream to defend the works against the leaping current, tumbling driftwood and floating ice. Eads had planned them carefully. Each breakwater, made of clustered piles, was in the form of an A, the apex about 200 feet above the pier site, the open ends widening to 200 feet apart below the pier. Slanting hinged aprons spread downward like wings from the outer piles toward the river bed. Sturdy bracing made the rough structure all but impregnable.[15]

The pneumatic caissons which William Nelson was building were described by James Eads as enormous elliptical bells, 82 feet long, open at the bottom, made of wood and covered with iron. The one for the East Pier was 60 feet wide, that for the West Pier 48 feet wide. A bottomless air chamber, 9 feet high, constituted the lower part of the caisson, the sand bed of the river being its floor until it reached bedrock. While it was slowly sinking to the sand, compressed air, pumped into it so that it could resist the heavy pressure of the stream about it, would keep the water pushed down from it. Upon its immensely strong roof the thick-sided hollow masonry foundation would be erected and later filled with concrete.

The East Pier caisson, the larger of the two then building, was provided with a false boat-bottom and launched on October 17, 1869, and on the following day it was towed to place in the river. The whole waterfront paused to stare and steamboats pulled aside as the ugly craft, noisy with the hammers still working on it, rocked and swayed over the water and entered the three-sided enclosure formed of widely spaced piles that rose 30 feet in the air. Outside of each end of the site had been moored one of the long barges bearing the 50-foot high hoisting frames, pumps, machinery, and accommodating the offices. Altogether there would be two dozen boats in use and fifteen hundred men employed. Atop the enclosing piles were ten giant screws which would be attached to the caisson to steady its descent in the river. Compressed air was pumped into the air chamber, the false boat-bottom loosened and towed away, and the open side of the stockade closed.[16]

It was a week after this launching, "a cold, raw day in October when the cornerstone of the Pier was laid; the wind whistled through the frames and network of wire ropes; the sky was black with huge volumes of smoke from the engines and boats on either side of the floating caisson; the din of hammers caulking the air-chamber below and riveting new plates above was deafening; the furious wind lashed the yellow Mississippi into something of an angry whiteness." The Chief Engineer, officials of the Bridge Company and a few others were ferried over the choppy water to the site. Chilled by the wind, but aglow over the occasion, they held their ceremony. It was brief, as the hammers were fairly held aloft for it. The visitors' boat bobbed away, a clang of metal, creaking of cables and shouts of men rising behind it.[17]

Barges of stone pulled up to the stockade and moored. The mechanical travelers on the workboats, operated by hydraulic jacks, began, under the orders of James Andrews, of Pittsburgh, contractor for the masonry work, to swing the stones to place. The outside of the bulging bell was covered with masonry borne on heavy supporting timbers. Then the stonework began to grow up from the stout roof of the air chamber and around the columnar caisson, which was being built continually upward, keeping well above water lest a sudden rise of the river swamp it. A stairway spiraled up through the main cylinder shaft, around which clustered seven air shafts.

Each day James Eads boarded the little tender, *Hewitt*, and went to the pier. After the cutting edge of the bell reached into the sand and the caisson rested upon the horizontal bearing timbers a few feet above the edge, he spent much time in the air chamber, where the excavators had gone to work. He had not forgotten the discomforts he had endured as a diver and he wanted to make sure how the men fared in the compressed air as they hacked at the sand and lifted large stones aside to a storage place on the girders. His air locks were unusual and a matter of intense satisfaction to him. Heretofore air locks had been placed abovewater, with large pipes leading down through the masonry to the chamber. But he had built his seven air locks *within* the chamber, about the main shaft where it pierced downward through the heavy roof and nearly to the sand. This arrangement was safer and more convenient.

His sand pumps, too, had proved excellent. They were simple, but of novel construction. Nothing like them had ever been used before. Their flexible hose reached down to nuzzle the stream bed, a high-pressure flow of water lifting and ejecting sand, gravel and even small stones through discharge pipes which extended up through the wrought-iron and timber ceiling to the surface of the river.[18]

Eads often made his way about the workboats, getting better acquainted with the men. Some of them were rough, indeed, with ugly tempers, and the Chief Engineer had been warned to carry a revolver and sheath knife, as was the unfailing custom along the waterfront. He did this, but no record is found that he ever had any difficulty. He joked, heard complaints, and managed to get along with the toughest of the workers. He held weight-lifting contests on the blacksmith boat, coming out second in them—slight as he was, his arms and shoulders were so muscular from the labor of his salvaging days that only the head smithy could outlift him.[19] And always he pictured to the men the Bridge as it would be someday, a mighty structure, the work of their hands.

Occasionally as he set out from the levee on the *Hewitt* he had guests with him. From the tender he would explain the spectacle of the huge pier being sunk without any dam to protect the site, and only a brick envelope strengthening the steel caisson as it reared above the water, waiting for its massive coating of stone. He smiled at their amazement when he told them that each of the six mechanical travelers on a workbarge could lift, by means of hydraulic jacks and rams, a seven-ton stone and carry it a hundred feet to place against the caisson.[20] The big bell that supported this heavy masonry had sunk far below the river bed, he said, and a visit to it would be a nerve-tingling experience.

After he had showed them about some of the workboats, he took them down the winding stairs and entered one of the air locks, which were vertical cylinders six or eight feet in diameter, where a door closed behind them and another door faced them. An air cock was opened into this narrow prison. When the pressure of the compressed air that flowed into it equaled the pressure in the air chamber, the door in front of them, which otherwise would have required a push of thirty thousand pounds to open it, swung back, and they stepped down on the sand floor of the chamber. They felt dizzy in the heavy atmosphere, their ears ached dully. The eerie light from the oil lamps, the hiss of escaping air, and recalled tales about the possible dangers in the tomb-like place added to the uneasiness. The sensation of having that ponderous masonry column suspended barely above their heads, and of the river tossing just outside the bell, was nightmarish. Their nervous laughter mocked back in unfamiliar squeaks. It was rare that some of the party did not have to flee the chamber almost upon entering it.[21]

From accounts spread by the visitors and the laborers a realization grew in the public mind that the work being carried on below the swift waters of the river was stranger and more colossal than anyone had dreamed it would be—anyone but James Eads. This man who had, wisely or foolhardily, dared to undertake it came to be regarded with a mingle of curious respect and suspicious doubt.

Day and night men hacked at the sand floor of the air chamber, men swung huge stones against the hollow column rising from the chamber roof, men riveted plates to the caisson growing abovewater or set the bricks of the 18-inch thick envelope. The Chief Engineer, pleased with the steady progress, was unmoved when public interest shifted from the bridge to a new ferry service, based at the north end of the city, which transported railroad cars across the stream. This ferriage was expensive, and there would be storms and ice tie-ups to interrupt it.

Extreme cold settled over the mid-valley in December of 1869. Ice cakes bumped each other in the wind-blown river, the ferries dared not budge from shore, trains dumped their freight on the banks and backed away. Again the pier drew all eyes. Bitter as the weather was, smoke puffed from the workboats, travelers swung their burdens. Even when the ice floes had grown so thick that a powerful tug, the *Little Giant,* had to be engaged to fight its way through them and carry the men back and forth, the work went on. Then for several days the men were stranded at the pier, the trip grown too dangerous. They slept on the *Hewitt* wherever they could find space, in blankets provided for such an emergency.

Unable to get to the pier, the Chief Engineer stood on the wind-swept bank looking through a telescope over the stretch of black water and drab ice to read large placards put up morning and evening at the site, bearing the superintendent's report. He swung the telescope anxiously upstream to where great ice cakes pushed up on the breakwater's inclined aprons and against a cable that reached to a superpile far upstream. Sawed in two, they slid down again and pushed on their way, held afar as they came opposite the pier. As soon as this ice run was over, James Eads resolved, he would have a telegraph line stretched from the pier to his town office.[22]

The day before Christmas of 1869, he insisted upon boarding the *Little Giant* for the pier. He wanted to visit the men stranded out there away from their families at holidaytime. Lunging ice cakes bore down on the tug, crashed against it, filling the air and deck with shattered crystals. There was a shout of welcome on the workboats, the men's spirits rose. The Captain knew how men felt being shut off out here. Things were not so bad, they told him, they were well fed on "Hotel de Hewitt," if not well bunked—some of them had rolled up on the boilers, some under them, some in the kitchen and the cook had raised a row about it. They were working, but not making much headway.[23]

Eads listened and chaffed, talked gravely to the superintendent and left. To those staring after him it looked as though the tug might never get to shore. Monster cakes rammed it, turning it from its course, driving it downstream, it righted, dodged, fought its way back. The Chief had come through that fury of ice to cheer them!

The ice danger subsided for a while, the men came back and forth until a few days after the first of the year (1870), when they were cut off from shore again. The *Little Giant,* trying to reach them, was beaten helplessly downstream. James Eads haunted the bank, scanning the bulletins. At any minute the ice might sweep suddenly forward, drawing half the river after it, throwing the weight of the surging flood against the pier. The hour came when he saw the jamming ice bear down toward the site as if to grind boats, stockade and masonry to bits, but the breakwater aprons caught the mass and shrugged it aside. A few days, and the stinging cold relaxed, the men came back to town as from a siege. The pier site rang and chugged again, the caisson sinking slowly with its burden.[24]

There had been no precedents for many features of the building of the Bridge, still the Chief Engineer had felt secure in the belief that engineering science would find a way. But as the East Pier sank far below the water he faced an emergency that the soundest engineering skill might not be able to meet: the ill effects upon the sand excavators of the increasing air pressure. A few of the men suffered occasional stomach pains and fleeting touches of paralysis, their breathing was somewhat labored. Uneasy about this, James Eads studied reports of discomforts experienced in other deep-water construction, but he found no inkling of extreme danger from air pressure—yet men had never worked at such a depth before.

The air chamber crew felt no alarm. They were amused that it was impossible to whistle in the sluggish air, and that a candle flame, blown out. would return again and again to the wick, they made light of their mild paralysis, even though it occurred oftener and lasted longer as the bell continued to sink. "A workman walking about with difficult step and a slight stoop was at first regarded as a fit object for jokes, and cases of paralysis soon became popularly known by the name of 'Grecian Bend.' The men had their own favorite panaceas, prime among them, 'Abolition Oil.' Galvanic bands of zinc and silver, worn around the wrists, arms, ankles, waist, and even under the soles of the feet, were believed to be a safeguard." The malady puzzled rather than frightened them. Some of them had worked in the air chamber ever since it reached sand and had escaped the ailment almost entirely, while others suffered from it before their first half day of work was over. It invariably attacked one as he left the air chamber, and a return to the pressure gave unfailing relief.[25]

When the chamber was 76 feet underwater a workman in it suffered such severe abdominal pains that he had to be taken to a hospital. Other serious cases soon developed. James Eads, now at grips with the human sacrifice that had too often been a part of great construction, strove harder to probe the mystery. He could get no advice from physicians, and had to study the malady at first hand. Superintendent McComas told him that the readiest victims had reported for duty ill clad, underfed, alcoholic, or all three, and that men on the night watches suffered the most, perhaps because of a lack of daytime sleep. With this little to go on, Eads ordered that only men in good physical condition be hired for the air chamber, that the length of the working day be cut down to three watches of two hours each, and that the night watch be abolished. As the malady waned, he hoped that little more would be heard of it. Light work, long rests and proper nourishment would likely conquer it.[26]

On February 28, 1870, the East Pier caisson reached bedrock, after four months of effort. The speed was noteworthy—it had taken the French engineers of the Royal Albert Bridge at Saltash, England, two years to establish a pier only one-fifth as massive and not nearly so deep. Fortunately, when the East Pier reached bedrock it was only 93½ feet underwater, although its high-water depth would be 26 feet greater. The men had suffered much discomfort, but there had been no fatalities. A large flag was flown over the workboats, smaller ones fluttered from the *Hewitt*. On land, cannon boomed in the still, cold day, setting off a bedlam of factory and steamboat whistles. St. Louis was jubilant, it had put its foot one stride ahead of the world, achieving the deepest underwater excavation ever made. Delighted citizens rolled out to Compton Hill to congratulate the Chief Engineer, telegrams poured in from everywhere. Doubtless James Eads was pleased with the attention showered upon him and his lately ridiculed project, but he appeared more excited over a piece of coral that he had found in a fragment of the limestone stream bed. There it lay in his hand, encrusted with crystals, a mute prehistoric record of the building of the valley by minute sea creatures aeons before the plunging river had begun to cut its bed in the stone.[27]

From this warming applause he turned to watch the severe labor of filling the hollow pier. First, the air chamber was filled until it was "reduced to the size of an Irishman," then when the last Irishman had crawled up out of it, the stairs were removed bit by bit, and the filling of the shaft commenced.[28] As though to make this more exhausting, the river began to rise. The cases of "bends" increased. Ten days after the freshet started, a new man who had worked only two hours came to the surface, remarked that he felt fine, toppled over and died. Death had come to the Bridge! James Eads was stunned, he had thought he was taking every precaution. He rushed his physician, Dr. Jaminet, to the pier to take charge of the men. While he was having a floating hospital fitted up, several more deaths from the caisson disease occurred. Aghast, he had the workday reduced to two two-hour watches, spaced between by long rests, and set up strict rules of relaxation and diet. But some of the toughest of the crew flouted the rules, eluded Dr. Jaminet's discipline, "sorely trying him," and landed on the hospital boat where they lay groaning while the Chief Engineer hunted out their families and cared for them. The day was reduced to three one-hour shifts, the rests between sternly enforced, and beef tea served during the rests. Complaints from the men grew bitter, much of the boredom with rules was drowned in the nearest shore saloon.[29]

At this time James Eads was diverted from his depression over the ailing sand excavators by a visit from Colonel Washington A. Roebling, who was to commence, in a few months, the building of the East River, or Brooklyn, suspension bridge. It had been designed by his father, the late John Roebling, who had also twice submitted designs to St. Louisans for a bridge across the Mississippi. Colonel Roebling had heard much of the unusual methods employed in sinking the piers of the St. Louis Bridge and wanted to examine them. Eads was glad to show the noted engineer about, he had patented all his own devices and had no objection to their being employed by other builders. He was particularly proud of the shaft of the West Pier caisson, where work was now well under way. This upper part of the caisson was a startling departure, being made of thick white pine staves, instead of metal, hooped together into cylinders eight feet long, the cylinders then placed one upon another and rabbeted together, growing upward as the bell descended, and enclosed in heavy brickwork against which the masonry of the pier was laid. Water had frequently penetrated the thick brick envelope of the East Pier and crept through the stairway shaft, but here it was held out by the staves, whose joints remained perfectly watertight without calking. The pine shaft was safer and cheaper than the iron shaft at the East Pier, Eads explained, and he was going to employ it at the East Abutment, which he had decided to sink to bedrock, where the depth was so great as to involve tremendous risks.[30]

The two builders enjoyed talking over their problems, not dreaming that the friendly occasion bore in it the seeds of a bitter feud. Much later James Eads was to accuse Roebling's aides of using his type of air lock and pine cylinders without royalty or credit. They, in turn, retorted that they and Eads had both obtained their ideas from early inventors. Eads thereupon produced descriptions of the early inventions to show where their designs left off and his began.[31] While the wrangle went on, the two great bridges grew above the streams.

There was a blithe air among the workers at the large East Pier. It was April, of 1870, the sky was blue, the subsided Mississippi danced in the sun, the hours were short and there was no end of coddling, especially of the air-chamber men. But Captain Eads had gone to New York on business, and at any minute bad luck might come out of hiding—disaster had often lurked around the corner for the Chief to get out of the way.

It was lurking now. A freshet began rampaging down the river, it surged about the pier, submerging the air chamber deeper than ever. The raised pressure beat on the men's ears and doubled their bodies in pain. The workday was cut to two 45-minute watches. Still the water rose until it was nineteen feet above the stone masonry, held out only by the sheet-iron caisson and its brick envelope which were being built frantically higher. The maddened current dashed savagely at the caisson, quivering it. On April 13, the strained envelope and sheet iron ruptured, water rushed in, cutting off the stairway. There was a dash to get the workers up the ladders of the small shafts, and in the excitement the air cocks and doors were left open. A diver had to be sent down each shaft to close its door so that compressed air could be pumped in to drive the water out.

When James Eads got back he found things in bad shape at the East Pier, and it would take a month of patching before the concreting could go on. Two weeks after the damage had been cleared away, the pier was completed, ready to receive its graceful, if heavy, masonry tower and, at the foot of the tower, the tips of two long steel arches. At the same time, West Pier, where the pine-stave cylinders held out the violent flood, reached bedrock 90 feet below the city *directrix,* nearly 98 feet below extreme high water, but in a submersion of about 78 feet. Now attention was focused mainly on the perilous sinking of the East Abutment where the limestone bed lay much farther down.[32]

For weeks at a time St. Louis would pin its pride, hopes and fears to the Bridge, setting up currents of boasts and sensational rumors. Word had got about that the caisson being constructed for the East Abutment would be a monstrosity, as big and ugly a thing as one could imagine. It would have to be sunk to an unbelievable depth—if it could be. Captain Eads might overreach himself this time. Then, all at once the Bridge was swept out of mind by the coming great steamboat race. The *Robert E. Lee* under the command of Captain John W. Cannon, was going to race Captain Thomas P. Leathers's fast *Natchez* from New Orleans to St. Louis.

The two boats left New Orleans on June 30, their progress flashed by telegraph to excited ports along the course. On July 4, the *Robert E. Lee,* stripped to a skeleton for the test, reached St. Louis six and a half hours ahead of the *Natchez,* which had done no stripping and had carried freight, loading and unloading at stops as usual. The St. Louis waterfront went wild. Three days, eighteen hours, thirty minutes from New Orleans! Captain Cannon of the victorious boat could hardly be got through the cheering crowds. James Eads, asked what he thought of the race, said that he was convinced that if the winning boat had been iron-hulled it would have made much faster time, perhaps five hours.[33] Iron boat hulls—he never let up on them! Just as though men had not got along with wooden boats for centuries! The big race over, gradually talk of it died down. The Bridge—when St. Louis failed of all other sensations, there was still the Bridge. Two sailing ships bound from Maine with granite to cover the piers above low water had been lost off the Florida coast on their way to New Orleans. That would mean another delay.[34]

The interested and the curious who swarmed to the launching of the East Abutment caisson on November 3, 1870, stared at the thing, incredulous. Reports had not done it justice. A wall of iron 30 feet high surrounding some irregular shape, moved down the ways by its own momentum, pushing up an angry wave as it struck the water. The enormous form rocked violently before it found a precarious balance on its false floating bottom, which had a draft of ten feet. What would Captain Eads think of next? As the top-heavy wall swayed, three steamers attached themselves to it and towed it away, having all they could do. The seven-mile trip from the shipyard to the site took five hours, and night had fallen by the time the caisson was anchored to position.[35]

Twelve more days were spent in finishing the caisson, removing its false bottom, and in assembling the last of the workboats and material, then the masons began their stone-laying, the night shift working by calcium lights. No sunken wrecks troubled the excavators here, but severe cold soon closed about the busy island of work. After Christmas of 1870, an ice gorge threatened even the breakwaters with destruction. James Eads left Compton Hill and its buzz of preparation for company and went to the abutment. There he set men to fortifying the breakwater with riprap— a mountain of ice might tear it away and overwhelm the monster caisson.

It was a day of sharp anxiety for the Chief Engineer, and the peak of peril not yet reached, for floes swarmed and jammed as far upstream as one could see. At dusk, cold and exhausted, he set out on a tug for home. "Why were the foundations so long in building?" he must have repeated cynically the question asked him countless times. With a small island of construction standing bravely in the midst of a violent stream, thundered at by spring flood and winter ice, with men manipulating ton-weight stones and men beneath the torrent gasping for breath, why did it take so long? Washington Roebling was finding that it would take him five years, at least, to build the foundations for the Brooklyn Bridge in the comparatively shallow water over the East River's granite bed. Why did colossal works cost so much time and effort? Only a builder knew.

For ten days more the ice battered the reinforced breakwater, erupting over it, threatening the abutment. Then suddenly it belched downstream, rocking and seething, sucking after it such a volume of water that the river seemed half emptied, and the workbarge *Allen* was tilted aground. With the sand pumps a basin was hollowed under it so that it could float level, the derricks and tall frames again erect.[36]

That would probably be the last of the winter's ice peril, the work could hum along. A mild few weeks early in the new year, 1871, brought visitors from near and far to the Bridge, engineers, tourists, financiers, curious to look over the world's deepest and most discussed underwater construction. James Eads took frank pride in showing them about. There were several new and unusual features at this abutment, he explained: an elevator to spare the air-chamber men the long climb up the spiral stairway, a lamp he had devised, its chimney connected with a discharge pipe that carried away all fumes, and a pleasant rest barracks fitted up for men off watch. How soon would the unprecedented foundation be completed? Now that everything was running with the precision of clockwork he could safely promise that a few weeks would see it through.

Even when the Mississippi began to rise suddenly in February, James Eads felt no alarm, so completely had his skill, at last, triumphed over the threat of floods. At the end of a fortnight's freshet the superintendent was able to report: "Everything in good shape and working well. . . . Immersion 101.1 feet." But on that very day, March 8, 1871, the storm elements, as if to offset Eads's mastery of the stream, struck out in fury. In midafternoon a tornado drove from the southwest with demoniac shrieks toward St. Louis, sucked its way above the city, dipped thirstily to the river and swept the east bank with a force that wrecked everything in its path. Large sycamore trees were torn apart, railroad trains whipped from their tracks, empty freight cars borne hundreds of yards through the air, their trucks dropping off in flight, a locomotive was tossed high over an embankment. The East Abutment works, directly in the storm's path, crumpled in a few seconds.

"In an instant the frames of the East Abutment were leveled with the water in one confused and shapeless mass," James Eads described the havoc later. "Hydraulic lifting machinery, air-pipes and hose, sand- and water-pipes . . . broken and carried down by the large timbers of the framework. These latter were mostly twelve inches square and from fifty to sixty feet long, and in falling were broken to atoms. The violence of the storm carried these frames with their top hamper on to the cabin in which were the air and pumping machinery and boilers. . . . Fortunately the boilers remained unbroken in spite of the tons of timber and iron suddenly thrown upon the slender roof above them, carrying it down in the general crash upon the engineers and firemen beneath. The men in the cabooses in the frames of the *Allen*, forty feet above the deck, went down with their cabins. . . . Strange to say, in the general ruin, with men almost as thick as bees in a hive, but one man was killed and eight wounded."[37]

The storm had done more than add another backset to the construction of the Bridge, it had left the public apathetic about the project and perplexed over the unshakable faith of its builder. There he was, tackling that hopeless pile of wreckage in a sweep of energy that rivaled the tornado itself, as though the late disaster and others yet to come would not mount the cost until the unfinished structure would have to be abandoned. James Eads felt the coldness and suspicion sharply and combated it with all his ingenuity. This was no time to let down. The movement of men and produce across the broad country more and more demanded the Bridge, and nothing, not even the mighty forces of nature, should forestall its completion.

The new frames went up with astonishing speed, and three weeks after the storm had swept the worksite into a chaos, the East Abutment stood like its fellows, an impregnable column firm on the stone bed, $127\frac{1}{2}$ feet below the *directrix*, 136 feet below extreme high water.[38]

CHAPTER IX

THE LONG STRIDES

THE four masonry foundations and their towers rose from the stone stream bed high above the water, but it had yet to be shown that metal arches could take the giant strides across them. The applause that pattered over the country was timid. James Eads understood its reserve—some men held faith in the Bridge because their need of it was urgent, a few because of scientific reasoning, but the mass of men would believe in it when it was completed and there was no remaining point upon which to hang a doubt. It made little difference, the Bridge would be there. The Chief Engineer now turned his full energy to the drama of raising the superstructure, the opening scenes of which had been played while the huge stones were being swung to place. Its serious theme was the difficulty of procuring the steel parts of which the arches were to be formed, melodrama and comedy were amply provided by the business antics of young Andrew Carnegie, one of the contractors.

For the world's first steel bridge James Eads had demanded metal of a quality and quantity never before required for any structure. Manufacturers in Europe and America had avidly sought the contract for making the arch parts, but after looking over Eads's specifications they admitted that they had no facilities for working steel in such large masses, nor men skilled enough for the exacting demands.

At last, on February 14, 1870, a year before the completion of the East Abutment, a contract for furnishing the parts and erecting the superstructure of the Bridge had been let to the Keystone

Bridge Company, of Pittsburgh and New York. James Eads could smile over this, for Keystone's president was J. H. Linville, who had earlier refused to imperil his reputation by sanctioning the "entirely unsafe and impracticable design." Linville was only lukewarm over this contract, having been persuaded into it by his energetic and voluble partner, Carnegie, who had grown interested in the Bridge by selling a large block of its bonds abroad at a handsome personal profit. In October, 1870, Keystone let a subcontract for the steel parts to the Butcher Steel Company, of Philadelphia, to whom Eads soon turned over the most minutely detailed drawings and descriptions.[1]

Each Bridge arch, James Eads showed, was to be composed of four ribs, each rib having two lines of tubes connected by a single system of bracing. A rib was formed by a series of straight tubes 18 inches in diameter and 10 to 12 feet long.[2] These tubes, each made up of six steel staves, would be joined by couplings, end to end, at a slight angle, the recurring faint angle shaping the curve of the arch. There would be 1,036 tubes, which meant more than 6,000 staves. There would be 1,024 main braces, 112 huge anchor bolts, besides tension rods, steel pins, nuts and so on.

The Butcher Company studied the details with growing dismay. The order was stupendous, the demanded quality superlative. The rigid test to which every fashioned part had to pass before it would be acceptable to the Chief Engineer of the Bridge was disheartening. The terms "elastic limit" and "modulus of elasticity," with which Eads had sprinkled the pages, were unfamiliar and terrifying. His furnished explanation that the elastic limit was that point to which a piece of material could be compressed or stretched under pressure or pull and still be able to resume its original form when the force was removed, and that the modulus of elasticity had to do with the ratio of stress to strain, failed to shed any glow of reassurance on the bewildered steelmakers.[3]

Andrew Carnegie flew into a dudgeon over the unusual requirements and wrote to the St. Louis Bridge Company: "This Bridge is one of a hundred to the Keystone Company—to Eads it is the grand work of a distinguished life. With all the pride of a mother for her first-born, he would bedeck his darling without much regard to his own or others' cost. Nothing that would and does please engineers is good enough for this work. . . . It is a new work emphatically. The very machinery to make the raw material has in large part to be created. . . . But he will come out all right. The personal magnetism of the man accounts for much of the disappointment. It is impossible for most men not to be won over to his views for the time at least. . . . You must keep Eads up to requiring only what is reasonable and in accordance with custom."[4]

Carnegie was soon in a tantrum over the iron parts too, some of which were to be made in his own ironworks (Carnegie and Kloman Company). The wrought-iron skewbacks, enormous plates and sockets weighing three and a half tons each, which, bolted to the piers, would receive the arch ends, were to be formed solid and the holes *drilled!* An unheard-of thing, but Eads insisted upon it. It was the high quality of the iron that pushed Carnegie the hardest, but he would produce it, he vowed, "if it cost as much as silver."[5]

These outbursts were understandable. The whole tradition of structural metalworking was being upheaved by the needs of this unprecedented bridge. And Eads did understand, but he could not recede—the quality of metal would have to advance. He had long ago had a material-testing machine devised by his assistant engineers, Flad and Pfeiffer, and set up in his St. Louis office, and had spent many absorbed hours trying the strength of metal, stone and hydraulic cement—now he saw to it that a testing machine, many of whose features he suggested himself, was installed in the Butcher steel plant. Through President Grant he borrowed Henry W. Fitch, Assistant Engineer of the Navy, and placed him at Butcher's to inspect and advise.

The work started propitiously, but despair soon took it over. Six months were consumed before a single steel stave—of the 6,000 and more that would form the rib tubes—had been found worth testing. Eads had Butcher's set aside their own carbon steel and try the product of a rival firm, the Chrome Steel Company, of New York. Jealousy and hurt were rife in the Butcher plant, and the ten thousand dollars paid them to make the change did little to soothe it. The chrome steel proved almost all right. A skilled metallurgist, brought from England, made a new mixture of it, but the plant manager modified it until the good results were lost. At last the homogeneous steel of another firm was adopted and proved excellent for the staves.

The problem of the staves was solved, but not that of the anchor bolts, each 34 feet long and weighing over 3,000 pounds, which were to secure the massive skewbacks to the piers. Most of these bolts were to be of steel, and of these four-fifths were breaking under test. Foremen and workers felt caught in a treadmill of trial and error. Whirlwinds of rebellion rose. James Eads did what he could to clear the air, but yielded not a jot of his standards. Throughout the summer of 1871 he divided his time between Philadelphia, Pittsburgh and New York, "his advice and ingenuity in constant demand." By August the testing machine was releasing many of the tortured parts unscathed.[6]

Believing that the steel production had been set on a straight path, Eads returned home. It was good to prowl his grounds in the golden September twilight with his townsmen, or to sit with them in the friendly drawing room, assuring them that the difficulties of supplying steel for the Bridge were being surmounted. To be sure, the piers were standing there in the river waiting for arch parts that should have been ready long ago, but the wait would soon be over and the arches strung up in short order. Every juncture and screw thread had been worked out in the plan details, and Keystone would have no trouble with their part of it.

While he was setting anxieties at rest, a gas flue leading to heaters in the Butcher plant exploded, and a month would be needed for repairs. Surely when this delay was over, the work would move right along. Only a matter of months now, Eads believed, and the Bridge would stand completed. It was well, perhaps, that the Chief Engineer could not foresee how unconcerned and leisurely Keystone would be in putting up the superstructure, or upon whose shoulders all threatening trouble would fall when Andrew Carnegie shrugged it off his own.

Sunday afternoons and evenings were especially gay and social at Compton Hill. If James Eads happened to be away, Eunice did the honors admirably. Interest in the Bridge filled the chairs in the drawing room and crowded the drives with surreys and carriages. The Sunday evening of October 8, 1871, brought an influx of late callers with an excited tidbit of news that had come over the telegraph wires soon after nine o'clock: A big fire was raging in Chicago. Ah, deplorable! ... heart-wringing! came comments of mock concern. It would be a pity if the railroad yards there should burn—its railroads had helped Chicago's population up to more than three hundred and sixty thousand, barely behind that of St. Louis. Chicago would doubtless stage a magnificent fire, it did everything on a pretentious scale. Pithy remarks about the fire, which, after all, might be a hoax, darted about, lapsed and revived each time the door swung open to the brisk autumn air and another caller.

On Monday, bulletins of the Chicago catastrophe shouted from every newspaper. The conflagration still raged on a wide front, driving all living things ahead of it. The summer had been hot and rainless there, the frame houses that made up a large section of the city were like tinder. Even the wooden paving blocks had burned under the running feet of the fugitives. St. Louis was agog over it, no one talked of anything else. Pity and horror were duly expressed, but antagonism toward the rival city was deepening, for Chicago had already begun to reap a reward from its disaster, its obscure name being publicized to the remotest corners of Europe. Dispatches of concern had come from far places, offering aid, and investment wealth promised to pour into the stricken city to rebuild it. "A searching test of our hearts was offered by the Chicago fire," a St. Louis chronicler wrote. "Of course we, with some display, sent money for the homeless, provisions for the hungry, and even resolutions of sympathy ... but privately everywhere could be heard, without unhappy tears, the pious scriptural ejaculation: 'Again the fire of heaven has fallen upon Sodom and Gomorrah; may it complete its divinely appointed work!' " Amidst the town's prevailing illusions, the Bridge, he declared, was "an anchor in an ocean of froth."[7]

St. Louisans reassured each other that Chicago's setback was undeniable, and that their own city could now more easily attain first rank in the nation. "The Future Great," they characterized it tersely. Only the completion of the Bridge was needed to make it simply "The Great," and James Eads's recent report showed that the contractors would manage this before long.

Into this complacency Messrs. Linville and Carnegie, on December 1, 1871, tossed a resounding bombshell—they were abrogating all Bridge contracts that had to do with steel because, they claimed, the adoption by Mr. Eads of an unspecified type of steel had violated their agreement. Oddly, it had taken them six months to become upset by this. St. Louis was dismayed. The ultimatum meant that Keystone would not build the arches, and who would? It was incredible that a reputable firm of constructors would abandon a work so abruptly.

Anxious to placate Mr. Carnegie, who could outpout a prima donna, the Bridge officials beset the Chief Engineer to soften his demands. He was adamant at first. The Bridge was not a makeshift thing, it was a timeless structure. Keystone, he thought, was too intent upon finding an easy way out of difficulties—and after all, was it for a lesser grade of metal or a money concession that they were angling? The St. Louis Company was paying the extra expenses; in all, they handed over a hundred and ten thousand dollars for adjustments. Moreover, they purchased six new lathes and loaned them to Keystone, and furnished Butcher's two additional boring mills. In the end, it did take a promise of more money to wreathe Mr. Carnegie in cheer so that the work might go on. James Eads was not pleased about any of this. A cloud of uneasy rumor arising from the incident was making the eastern investors, whom he had worked hard to attract, so restless that something must be done to avert a crisis.

THE LONG STRIDES

A meeting of all interests was called in New York. The representatives assembled nervously. The atmosphere was explosive even before Anderw Carnegie arrived, crackling with ideas. He bobbed up at the first opportunity with a resolution that the services of an engineer of *experience* be appointed and given authority (along with Mr. Eads, perhaps) "to alter, amend or curtail the existing Bridge plans so as to insure the completion of a satisfactory structure at the earliest moment and at the least possible expense." His choice for the post was really not himself, it was his partner, Linville.

The assembly came taut. The move was a bold attempt to push the Chief Engineer into an impotent position and thereby relieve Keystone of all irksome standards. Every eye turned upon the designer of the Bridge—the Captain had a hot temper at times, how would he take this? James Eads could usually be counted upon to do the unexpected, and he did it now. Rising in his place, outwardly unruffled, he said that he was in favor of adopting Mr. Carnegie's resolution, for he thought that there should be a thorough examination of the Bridge plans and so-far structure. His manner was as gracious and cordial as though he were inviting friends to a chess tournament.

There was surprised whispering, the stockholders were disarmed, a destructive wrangle averted. The resolution, coolly received when it was offered, was now unanimously adopted. It was impossible to tell whether Carnegie or Eads had triumphed. However, an attempt to edge Linville in as examining engineer was firmly parried. The new post went to James Laurie, a Pennsylvanian, who was to report at St. Louis on March 1.[8]

From the ferryboat as he crossed the Mississippi on his way home, James Eads could see the masonry towers for the Bridge looming one by one in the winter mist. Only a short while now, and transcontinental traffic would move along tracks stretched through the tops of the arches. The ferries were on their last miles, they would vanish over the horizon of time as had the Conestoga wagons and poled keelboats. This was the age of steam trains and steamboats—after them, who knew?

It was the day before Christmas. The wharves were full of song and bluster, saloons and billiard rooms boisterous. Late shoppers hurried about the cold streets, carriages piled with bundles rolled homeward in the dusk, firelight twinkled past gaily ornamented trees and glowed through unshaded windows. The peal of church bells followed James Eads out of town after all other sound but the jingle of harness and hoofbeats of his carriage horses had died away. There was a flutter of greetings as he alighted at the side door, then his own cheerful fireside and a forgetfulness of strain.

The holiday season was filled to the brim for him. He was a considerable personage, and apparently far from irked with his role. He graced formal receptions and balls and was a familiar figure at the theater, sitting stoical through tragedy or laughing his fill at comedy. His speeches vibrated with dignified fervor in public halls and at banquet tables, the subjects, whatever they might be, tacking around to transportation and commerce. Commerce, he held, was the paramount issue among men. "Commerce with her breath of steam, commerce with her nerves of lightning!" he had once exalted it; "Commerce, whose voice controls grim-visaged war. Commerce, the ruler of kings! Commerce, whose ships gather her tributes from the uttermost parts of the world! At whose command the depths of the sea and the bowels of the earth yield up their treasures. Commerce, the annihilator of space, the builder of cities, the founder of empires. Commerce, who blesses the broad earth with Peace, with Plenty, and with Happiness. . . . This is the monarch who rules the world now."[9] If this tribute was grandiose, it was because there flowed behind it, in Eads's mind, the mighty, regal Mississippi, the God-given free road to market.

Wherever he went James Eads was pointed out as the man who had built the huge piers that stuck up out of the river ready for arches—arches that might never exist, some added. At any rate, the fame of the piers and their deep construction was farflung, tourists were attracted to St. Louis especially to see them, and to catch, perhaps, a glimpse of the daring man who had set them in the tempestuous stream.

Among the visiting notables of the winter, 1871-72, who took a keen interest in the Bridge and its designer was Europe's Beau Brummell, the Grand Duke Alexis, son of Czar Alexander II. With his suite, Alexis had traveled across the country on a special train, leaving "national conquest in his wake." Great preparations had been made for him in St. Louis, where he arrived from Chicago on Saturday morning, January 6, 1872. Mounted police met the ferryboat and escorted the royal visitor through crowds to the Southern Hotel where other police would hover him, for the city authorities had received a number of threats against his life. At at reception which followed a welcome by Mayor Brown, the handsome young Grand Duke, splendid in his colorful uniform with its elaborate insignia, and the slender, arresting bridge-builder in his correctly formal attire, shared the limelight.[10]

After refreshments had been served, a "quiet drive" over the city was begun. Two thousand persons lined the streets to see the Grand Duke enter a "four-horse turnout" that had been made gorgeous with gold trimmings. The carriages of the principal guests fell in behind as the cheers of the spectators and the sharp clack of ironshod hoofs mingled in the wintry air. As they rolled out the high streets near the river, the piers that would someday (it was hoped) support the world's greatest bridge swung around the horizon and marched majestically away in the grayness. In the evening the visitor was treated to a theater party at the Olympic, where Lydia Thompson played in a burlesque performance of *Bluebeard*.[11] One can almost see James Eads, the once salvage diver, sitting not far from the royal guest, never forgetting his dignity as he laughed at the nonsensical play "till he was red in the face."

On Monday, besides presenting a diamond bracelet to Lydia Thompson, the actress, Alexis, with his party, was escorted by the Mayor and a number of "representative citizens" at one o'clock to the Merchants' Exchange on Main Street. St. Louis was very proud of the beautiful interior of this building, with its walls decorated in historical scenes painted by local artists, but Alexis appeared unimpressed—he did not hesitate to show boredom if he felt it. His interest picked up when they left the Exchange for James Eads's office, also on Main Street, not far away. There he spent a half hour chatting, asking questions about the Bridge and looking over plans which his host explained to him. He was enjoying himself thoroughly now—he was more thoughtful than his light social touch advertised. Laughingly he suggested that a similar bridge might be stretched over the Volga and a tunnel dug under the Dnieper with American initiative and Russian nerve.[12]

That evening James Eads, taking his second oldest stepdaughter, Adelaide, petite, blond, fond of a good time, rode through a snowstorm to a ball and banquet given for the Grand Duke at the Southern Hotel. It was after ten o'clock when the Arsenal Military Band struck up a march, and Alexis was ushered to his place in front of the "Arbor of Love," a flower-bedecked bower, where he greeted the stream of guests that passed him.[13] If all this pageantry was dreamlike to James Eads that night it was because he had stepped down into it from a great reality that still held him rapt—he had been preparing a lecture on "The Theory of Light," to be delivered the following week at the St. Louis Academy of Science. Light! Where did light leave off and other phenomena begin? All this color, warmth and sound, were they slowed light, or slower kin of light? What relation did light have to the human figures, or even to the very souls, about him?

A quadrille was forming. Adelaide, with General Hatch, was being bowed to place opposite Alexis and the Mayor's wife. The next dance was the lancers quadrille. James Eads and a Miss Formel were placed opposite Alexis. The Grand Duke was an expert dancer, and the Captain was not doing so badly himself. They bowed and turned, pranced forward and back, wove in grand right and left. It was all very exhilarating. After a gallop and waltz, the doors were flung open to a banquet hall and supper announced. It took protocol and deftness to form the lines into the lavishly decorated room. At the head of the royal table stood two emblems, a white bear holding Russian and American flags, and a Goddess of Liberty set off with the motto: "May Russia and America ever stand united."

Two days later the Grand Duke, laden with gifts for the Indians, was off in his private train to be met at Kansas City by General Sheridan and conducted on a buffalo hunt by "Buffalo Bill" Cody.[14] And James Eads was submerged again in his study of light.

His talk, given on January 15, was an inaugural address, for he was being installed as the president of the St. Louis Academy of Science. The subject so filled and elated him that he did not get around to a mention of the Mississippi. He explained the comparative wave lengths of light, sound, color and heat, the familiar spectrum, the invisible rays (ultraviolet and infrared) beyond the spectrum. He described minutely the use of the spectroscope in separating luminous vibrations of gases and metals, identifying and bounding the so-far known elements of which the universe is made. As he talked he was, for the most part, the impersonal scientist, yet, at moments, a mystic looking beyond the known into the unfathomed.

"Some of the results of scientific inquiry have been so amazing," he said, "that the powers of the human mind seem almost elevated by them to the verge of the omnipotent."[15] He tasted of an ecstasy, and his listeners were borne into it with him. It was always something of an emotional experience to follow James Eads in the happy vagabondage of his reaching mind.

The Bridge, the Bridge—when would the Bridge be completed? The query challenged James Eads on every hand. The answer lay with the steelmakers, and some of it had been given out: "Rolling staves is a very simple matter now," they reported. Presently the tubes could be made up. Still the couplings to connect the tube ends were lacking, although work had started upon them more than a year before. Eads faced the dreary fact that he must again take up his pleading, pacifying and subtle driving.

In March, 1872, James Laurie, appointed to examine the Bridge and suggest improvements in the superstructure's design, arrived in St. Louis. James Eads received him cordially, offered him every facility for his task, and showed himself so cryptically suave throughout the ordeal of having his plans picked over critically that one cannot be sure whether he was glazed in diplomacy or confident that the examiner had come on a futile mission. Laurie, a serious, thoroughgoing Scotsman, plodded through the voluminous drawings, specifications, contracts and bills turned over to him. He spent days inspecting the foundations, he ignored the sly teasing of Eads's aides, who resented the intrusion heartily. And in April, satisfied that nothing had escaped his scrutiny, he reported that he had little or no criticism to offer. However, he suggested in a halfhearted way that a change in the planned approaches would save a few thousand dollars, knowing full well that the St. Louisans would not thus mar the beauty of a structure costing millions. His services were at an end, and Andrew Carnegie, in blustery chagrin, faced the same rigid demands as before.[16]

Still the Bridge foundations waited. No couplings were forthcoming. Steel couplings of such size could not be rolled, Keystone told him, and the attempt to do it would be ruinous. So far, Butcher's every trial had failed. James Eads went to Philadelphia and took up his old stand in the steel plant, and there he toiled most of the summer of 1872. Change after change was made in the rolls, and by September half the slight output survived the testing machine. Convinced that the work would go on nicely without him, Eads left for England on a financial mission. When he returned six weeks later he found that the couplings had suffered another slump. He accepted the ones that bore up under test, offered a liberal payment to the steel company to spur them on, and went to St. Louis.

Across the country came racing after him the unexpected ire of Andrew Carnegie—Eads had gone around him, that extra payment to the steel company did not trickle into the pockets of Keystone! Ignoring the fact that he had once thrown over all contracts because Eads had insisted upon high-quality parts,

THE ST. LOUIS BRIDGE

Picture at top shows erection of the West and Center arches, view looking northeast. *Photos from the St. Louis Historical Collection of Dr. William G. Swekosky.*

Carnegie now bounced with indignation because the Chief Engineer had accepted "inferior parts." He declared that the Keystone contract was totally invalidated and that the whole risk of erecting the arches must devolve squarely upon Eads and his associates.

Rocked with a sense of outrage, the St. Louis company ached to get rid of Carnegie, but that wily gentleman had an intimate connection with railroads, and the good will of railroads was important to the Bridge. They said things that eased their minds, and turned the matter over to the Chief Engineer. He was not in a tender mood himself, wasted months had slipped by, he had spent the best of two summers in the steel plant performing the neglected duties of the contractors, he had been punched from both sides by the pugnacious Mr. Carnegie. Holding his temper in leash, he wrote curtly to Linville that if he objected to the use of steel parts that did not measure up to his standard, it was "imperatively incumbent" upon him to secure some that did. "I have, however," he concluded, "no authority to release you, as you requested, from the responsibility you have assumed."

Linville and Carnegie were dashed. There was something peculiarly stupid about Eads at moments, he had failed completely to grasp the underlying subtlety in their complaint. They would be more explicit. They sent him a copy of a resolution that ended: "It is believed that a plan may be devised by which, at an additional expenditure, it would be safe to use cast-steel couplings . . . provided that the St. Louis Company deem it advisable to pay the additional expenditure."[17]

It was all quite clear to James Eads now—Keystone's hand outstretched, palm up, again! This time he would ignore it. After a while a telegram came from Linville saying that he would proceed with the work until he found it unsafe, but Eads and his associates would have to assume all risks of erection. It seemed a smug threat. Eads felt goaded, tormented. There sat the masonry foundations in the river, waiting, while a cynical public chaffed about them, while railroads wondered whether it would not be well for them to reach the Mississippi at a more northerly point, and hungry interests ate into the Bridge funds. With two members of his company, Eads hurried to Philadelphia to meet Carnegie. There they soon had to admit to each other that Carnegie was in a position to hold up the arch construction ruinously, and that there was nothing they could do but pay him the bonus of twenty-five thousand dollars that he wanted for cast, unrolled steel couplings, or somewhat less for iron couplings, if it came to that. Besides this, a bonus of thirty thousand dollars would have to be paid for reasonably prompt deliveries.

Now that these fat sums had been wrung out, the parts were made faster, but not furnished at St. Louis faster. They were faulty, was the excuse. Keystone's superintendent, Colonel Piper, wrote: "I am certain that the Bridge is a-going to the bottom of the Mississippi." Like a hurricane James Eads descended upon Keystone. The matter of couplings had consumed more than two years, he blazed, and only half the needed number produced. Butcher's had done quite well with them for a while, a way having been worked out to cast each coupling in halves. The molds were "swept up" in dry sand, and when the ingot was removed from the molds there was no pulling or cracks. They had been heated, rolled, tested and sent here to Pittsburgh, where they broke under hammer strokes. The test called for several hammer strokes, but not for a beating with a 23-pound sledge hammer, Eads pointed out. The tests should be more in line with the strains the parts would endure in the Bridge—why submit them to arbitrary destructive treatment? Arguments were wasted. Eads saw that steel couplings would not be furnished, and worked intently to redesign certain factors of the Bridge so as to shift strain from the iron couplings that would be substituted and which Andrew Carnegie's ironworks would fashion. The more fatigued he felt over this crisis, the plumper, rosier and more complacent Andrew Carnegie seemed to grow.[18]

There had been a few diverting incidents for James Eads during this tense year, 1872: the University of Missouri had conferred a degree of LL.D. upon him; the National Academy of Sciences had elected him a member; and the Chief of the U.S. Navy Engineers had asked him to devise further improvements in warship gun carriages, which he had done, taking out his nineteenth patent on naval ordnance and its manipulation.[19] But in all, it was not a placid period for him. During the recent presidential campaign he had had to choose between two of his friends, General Grant, on the Radical Republican ticket, and Gratz Brown, who had shifted back to the ranks of the Liberal Republicans and been given second place on a ticket headed by Horace Greeley. It was a hot contest, even schoolboys taunted each other with doggerel insulting one candidate or the other. Both Republican factions in Missouri appealed to Eads: Would he use his influence for this one or that one, write a letter or speak a few words that might sway certain critical opinion?[20]

James Eads appears never to have swerved from his position as a Liberal, and he declared now for Gratz Brown. It was a hard stand to take, but Grant was not to hold it against him in years to come, years warm with mutual affection, though stinging with controversy. Greeley and Brown went down in defeat, some thought because the Democrats had given them the kiss of death by nominating them, too. Gratz Brown had not helped their cause by making a speech at a Yale class banquet after his wineglass had been filled too many times and his low estimate of Easterners just would out. Greeley had died soon after the election and Gratz Brown had retired into a political gloom that would never lift.[21] Besides these depressing incidents, Eads's quarrel with Washington Roebling, builder of the Brooklyn Bridge, reached its height. Roebling made "severe personal remarks" about Eads, who grew severely personal himself, ridiculing his antagonist with a facile pen.[22] This quarrel, however, was to fade out, the principals too engrossed with their mammoth works to keep it going.

It was spring of another year, 1873, when James Eads saw the first steps taken in the erection of the Bridge arches. No false works would be set up in the river, lest navigation be impeded. Instead, temporary frame towers were erected on top of the masonry towers that rose from the foundations, and cables reaching from these would support arch halves, begun at the same time by securing enormous steel anchor bolts to two foundations at the foot of the permanent towers. The semiarches would grow toward each other and meet over the stream midway of the span. The fashioning of the cables, commenced months before, had drawn its share of public curiosity. Made up in sections 27 feet long, of steel bars 1 inch thick and 6 or 7 inches wide, "The putting together of a cable four hundred and ninety feet long was no small matter," and the cable itself no frail support. This was the first time that cantilevering, or bracketing, had ever been used extensively in building a bridge, and ranked along with the sinking of the deep piers as one of "the most difficult problems ever attempted by an engineer." Other engineers eyed the experiment dubiously. Surely James Eads was an original and daring man.[23]

As Keystone was making fair time with the construction, Eads thought he might be spared for an urgent business trip. No sooner had he got out of sight than the work lagged, then halted. When he got back three weeks later, it was taking a peaceful siesta in the June sunlight. In a fury he went after Keystone. Hot as his temper was, when roused, Eads rarely lost his self-possession, keeping a crisp dignity as his words whittled neatly to the point. Linville felt aggrieved at his sharp criticism—he could not be expected to go on with the construction when there were no more coupling bolts for the tubes, could he? Eads reminded him that Keystone had had the details and material for the bolts nearly two years, and that the entire lot could have been manufactured in any two weeks. Linville admitted this, but the fact remained that he just had not got around to it yet. This set Eads off on a gorgeous rage, which reached in every direction to other deficiencies.[24]

Despite his outburst, the leisurely air of the contractors remained undisturbed. Creepingly the halves of each arch, supported by the cables, stretched out from the piers, while Linville and Carnegie wondered honestly, if not strenuously, how two halves could be joined into a complete arch—or *if* they could. This confounded puzzle had been concocted by the Bridge Company's

chief engineer, whom Carnegie later described as "an original genius *minus* scientific knowledge to guide his erratic ideas of things mechanical," and whom Superintendent Piper of Keystone had demoted from a respectful "Colonel Eads" to "d——— Jim Eads."[25] The difficulty in connecting the semiarches lay in the fact that Eads, in his eternal figuring, had found it necessary to have them made a bit longer than the distance which they were to span, since the weight of the structure would compress them, once the supporting cables were removed.

"The ratio by which we have to multiply all dimensions for construction is 1.000363," James Eads had written in his specifications. Therefore every tube had to be made a minute fraction too long and the semiarches would lap at the middle by 2.256 inches. As though this lap had not been bugaboo enough to the contractors, Eads later had announced that because the steel modulus of elasticity had proved too low, the lap of the half arches would be 3.252 inches! At last Keystone decided to dodge this troublesome matter—when the time came for inserting the final tubes, their foremen would retire from the scene, leaving the problem with the St. Louis company.[26]

Engineers generally agreed that a 3¼-inch lap in 500 feet was serious and that any attempt to join the semiarches would be fraught with peril to the structure. Eads, for a good reason, was not alarmed by that specter "Impossible" which was stalking about again. He had spent some of his "original genius" devising a way to close the arches and was comfortably waiting for the time to use it. But he was disturbed about the Bridge Company's financial straits, and he was ill, having just returned from the Mississippi delta "with a severe cough and bleeding lungs." His doctors insisted that he take a rest at sea where he could hear nothing of the Bridge for a while, or he might not live to see it completed.[27]

Leaving his assistant, Henry Flad, in charge he went first to Philadelphia, where some of the harassed St. Louis company were trying to convince Linville and Carnegie, who met them there, that it was urgent to push the work without further dallying. These pleas made no impression. It took a bonus offer of thirty-five thousand dollars if the arches were closed by January 1, 1874, and of thirty thousand more if the Bridge railroad tracks and highway were ready for use by the following March 1, to bring the Keystone partners to life. With encouraging animation they signed the agreement. Eads, armed with these bought promises to dangle before English bankers for a much-needed loan, sailed from New York, August 20, 1873.[28]

Glibly the heads of Keystone had agreed to join the arch halves by the next New Year, although sincerely doubting that they could do it by kingdom come. And since they could not earn the alluring bonuses, they bent their wits upon how to secure the greater part of them anyway. They sent a blunt notice to the St. Louis company that they would not execute the contract they had signed the day before, but would substitute another one for consideration. Five days after James Eads sailed for England, Messrs. Linville, Carnegie and Piper arrived in St. Louis with their new proposition. It was not greatly different, they said. The bonuses and all were in it, but with a difference: the thirty-five thousand dollars for closing the arches by January must be paid in installments, beginning at once—which would, of course, leave it utterly ineffectual. All that Keystone had meant to guarantee, they said, was that the arches would be *ready* to close by January. As for inserting the final tubes, they would not promise to do that by any time.

The St. Louisans felt caught, their arms pinned down. They could not afford a contractors' strike at this critical time, with the Chief Engineer on his way to convince European capital that the Bridge would soon be completed. Moreover, Carnegie's clamorous insistence wore them down. At last they agreed that the arch bonus would be paid "irrespective of the date of closing them," provided that Keystone would at once furnish an adequate force of men and all necessary apparatus.[29]

At this propitious moment, with James Eads out of the way for a while, an attack upon the Bridge was made from an unexpected quarter. Although for five years the much-pictured and published plans had brought no protest from rivermen, when the semiarches began to reach far out over the channel some of the steamboat interests became uneasy. Everything dire had happened to steamboating, a shoaled and cluttered channel, the competition of railroads, and now this Bridge, with its arch middles hardly 55 feet above the river at high water. The smokestacks of some of the new packets reared nearly twice that high—those of the *Great Republic* 105 feet, those of the *Ruth* 107 feet. The Keokuk Steamboat Line had made a formal protest to Secretary of War William Worth Belknap, whom its officials knew personally, against the completion of the Bridge, and on the very day that James Eads had sailed for England an order was issued from Washington convening a board of engineers at St. Louis to examine the Bridge and report upon it.[30]

With no warning to the Bridgebuilders, this board convened on September 2, 1873. The Bridge Company, hearing of the meeting, sent a note to the convention, saying: "This company has been in existence and practically at work for over five years; it has expended or become liable for about $9,000,000;[31] its plans for the Bridge have been published and circulated widely. . . . During the whole of this period no complaints have been made by either the government or the people. . . . If therefore . . . this company learns for the first time that, just as this work is on the eve of completion, your honorable board is convened for the purpose of examining the construction of the Bridge . . . it cannot but be somewhat startled at the intelligence." They begged that they might send some of their officials to the meetings to learn the grounds of complaint, and counsel to advise them.

The board of engineers, sitting with their own attorneys at their elbows, curtly denied the Bridge Company the right to have counsel present, but gave grudging permission for a few Bridge officials to drop in and look on, making it clear that the looking was to be done in silence. "Such serious obstructions to navigation as we may find, if any, and such modifications, if any, as we may propose," the amazing reply went on, "will be based upon our own determination of facts."

For two days the board sat in solemn session, hearing the charges preferred against the Bridge by the Keokuk Steamboat Line. These complainants, armed with lengthy documents signed by rivermen, testified to the absolute necessity of supertall smokestacks. And that even if these were not too high to pass under the arches, the difficulty of steering a boat through the overwide spans, with all the piers too far away to serve as guides, would be enormous. One of the Bridge officials rose to refute the testimony but was promptly ordered down, steamboatmen whom they brought in for witnesses also were hushed. Only during the last few minutes of the last session were the Bridge Company witnesses allowed a few hasty words: chimneys were often ridiculously high, these steamboatmen admitted, and even then, they could be easily hinged to lower and raise. It would be far more sensible, they thought, to modify steamboat stacks than to change the Bridge. As for the much dwelt-upon difficulty of piloting a boat through spans of *too great width,* it was purely fanciful.

To this hurried defense, Major Warren, of the engineers' board, replied in sweeping dramatics: "If a thousand steamboat men should come and say that this Bridge is no obstruction, it could not change my opinion!"[32]

It was all over. The board's report was dispatched to General A. A. Humphreys, Chief of the United States Army Engineers. It denounced sharply the Bridge's low arches and the wide span through which a boat could hardly find its way. The unusual document admitted that there would be more danger of a boat's striking piers that were set closer together, but reminded that in such a case "the damage would be to the hull alone, and even if so great . . . as to sink the boat, time will generally be afforded to save the lives of the crew and passengers."[33]

The report ended in a high climax: "The Board have very carefully considered the various plans proposed for changing the present structure, but find none of them satisfactory. . . . They would therefore recommend that a canal . . . be formed behind the East Abutment of the Bridge." This canal would be eight hundred feet long and spanned by a drawbridge. Any remedy adopted was to be the responsibility of James Eads and the Bridge Company to carry through.[34]

Chief of Army Engineers Humphreys read the report with entire agreement and sent it to Secretary of War Belknap, recommending that it be submitted to Congress for action.[35]

The St. Louis company talked themselves into the belief that the whole affair could be dismissed as a farcical piece of spite which would come to nothing, and gave their attention to hastening work on the superstructure, two arches of which were now ready to close. Chief Eads had sent word from London that a loan of a half million dollars from the House of Morgan depended upon the joining of at least one set of arch ribs by September 19. He knew that there need be no trouble in completing the arches, it would take only a few hours to connect a rib with the device he had left ready for use. This method consisted of having duplicates made of the final tubes, of cutting these duplicates in half, removing five inches of their length and running screw threads inside of them. He had had short wrought-iron cylinders made with corresponding screw threads, and this internal plug would be screwed far enough into the tubes (turned by powerful levers thrust in holes provided for them) to shorten the tubes until they met, then later unscrewed to the desired permanent length. Steel bands would cover the exposed part of the plug.[36]

Henry Flad, in charge during the absence of the Chief Engineer, had been positive all along that this device would not be needed, insisting that the dangling semiarch ends could be jacked up until the increased arc did away with the lap. Eads had agreed that he might try this up to the point where the cables bore a certain strain, twelve thousand pounds to the square inch. It would not, however, be successful unless the weather were cold enough to contract the ribs slightly, and cramped finances could not wait upon the thermometer.

Flad set the men to jacking up the cables of the west arch ribs. On Sunday, September 14, 1873, five days before the London deadline, one of the final tubes—not of those which Eads had made extensible—was hoisted. The weather was fair and growing warm, the arch halves steadily expanding. Tug as they might, the men could not make the connecting tube do anything but lap. On Monday the work started at dawn, but even at that early hour the space for the tube was too short. Flad thought he might wait a day or so for cooler weather, but found that a hot wave was moving up the river from the south. He decided to cool the ribs with ice—it should be a simple matter to shrink out a few inches in five hundred feet of metal. He had long troughs built under the ribs by the too few men now furnished by Keystone, and thirty thousand pounds of ice brought to the site. At two o'clock on the morning of the 16th the men began wrapping the ribs in gunny cloth which soaked up the ice water kept running in the troughs. Once for a short time the half ribs were only five-eighths of an inch too long. Success teased, promised, eluded the intent workers.

The sun rose with the breath of a furnace. At midday it was eighty degrees in the shade down in the city, but there was no shade on the Bridge top. The men labored, the heat beating down on them from a merciless shimmering sky and steaming up from the glassy water. At five o'clock in the afternoon the temperature had risen to ninety-eight degrees, and the tubes had grown hopelessly long.

All night the work went on, the tired men stumbling until Inspector Cooper feared they would fall in the river. A warm wind persisted from the south. The morning of the 17th was stifling. Sixty tons of ice had been used, some of the men had been on duty for fifty hours without rest, the tubes still lapped, and the London deadline was only two days off. Henry Flad, whose sound enough theory had been beaten by the unseasonable weather, gave an order to use the extensible links left by the Chief Engineer. By nine o'clock that night two arch ribs had been joined.[87]

On the following morning Flad cabled James Eads in care of Junius Morgan, in London: "Arch closed. Last tube at nine yesterday." Mr. Morgan opened the dispatch, for Eads, the only one who had not worried over the closing of the arches, had gone complacently to Paris. When word reached him there that all was well and the loan had been saved, he sent a message of congratulation to Henry Flad, concluding it with his well-worn motto: "Disasters and serious accidents are always evidence of bad engineering."[38]

It was the last week in October when James Eads returned to St. Louis. The Bridge arches now all but spanned the river, yet the company was steeped in dejection, the workmen were depressed, the public cynical. Chief Humphreys of the Army Engineers had spread the word that Secretary of War Belknap agreed with him and his recently busy board of engineers in condemning the Bridge. The grand structure was going to be mutilated, or made ridiculous by a costly canal around it.[39]

Astounded that the Bridge had been tried and convicted, and no defense of it allowed, Eads started for Washington, taking with him Dr. William Taussig, vice-president and general manager of the Bridge Company. Erstwhile chemist, physician, politician and banker, this shrewd and efficient native of Bohemia was a neighbor of James Eads, and his right-hand man in many a Bridge crisis. At the Capital they tackled General Humphreys at once about the highhanded actions of the board of engineers and his agreement with them in condemning the Bridge. Humphreys patted the whole thing down, it was nothing to get upset about. His remark that Secretary of War Belknap stood back of him in the matter? Well, that was a mere inadvertence on his part. Actually General Belknap had taken no stand in it.

Within a few hours Eads and Taussig had found that the board of engineers' report had gone to Congress, its threats of mutilation of the Bridge approved by Belknap. Disgusted with the chicanery dealt out to them, and alarmed over the possible fate of the Bridge that had cost so much in time, money and human lives, James Eads and Dr. Taussig debated what they might do next. They could appeal to President Grant for aid, but it would not be a comfortable thing to do: Eads had bent his influence toward the election of Grant's opponent, Horace Greeley, and Dr. Taussig was the very county judge who had cast the deciding vote that had barred the then penniless Captain Grant from the post of county surveyor. But this was no time to quibble over the past.[40]

It was a sultry Washington morning when the two tense, determined St. Louisans, in meticulous dress, presented themselves at the White House. Reporting the scene years afterwards, Dr. Taussig said that the President met Captain Eads "with outstretched hands, greeting him warmly." He was slightly less cordial to Taussig, addressing him as "Judge" to show that he had forgotten nothing, then assuring him that he held no rancor at being denied the job of country surveyor since his present one was more attractive. Grant listened in serious surprise to his callers' story, he had never heard of the fabulous report of Belknap's board of engineers. He had the Secretary of War sent in and plied him with questions: Did the St. Louis Bridge conform to the demands of Congress? Hesitantly Belknap admitted that it did. Had the former Secretary of War approved the structure? Well, yes, he had. But, Belknap added defiantly, he was acting within the authority of his office, which included even the right to have the Bridge torn down if it obstructed navigation.

Anger flared into Grant's eyes, his broad jaw set. "You cannot remove this structure on your own judgment," he retorted. "And if Congress were to order its removal it would have to pay for it. It would hardly do that to save high smoke-stacks from being lowered when passing under the Bridge. If your Keokuk friends feel aggrieved, let them sue the Bridge for damages. I think, General, you had better drop the case."[41]

The Bridge was saved. The petitioners thanked the still irate President and left for home with the good news. Soon after that, President Grant came to St. Louis and called on James Eads at his office, talked with him awhile and went with him and Henry Flad to walk the plankway that had been laid from arch to arch. It was a heady adventure, even for those accustomed to it, but

Grant trudged fearlessly along, taking in everything with sharp interest. When the party returned to the office, a bottle of brandy and a box of cigars were set out and they "all sat down to a draughtsman's deal table."[42]

Before this visit, while still seething over the latest attack upon the Bridge, James Eads had begun a review of the board of engineers' report, addressing it to the company. At his desk in the library on Compton Hill, with the stripped tree branches swaying past the windowpanes and a wood fire leaping on the hearth, he picked that critical document to bits. With relish he challenged the testimony that the board had accepted. In ruthless sacrilege he blasphemed the sacred loftiness of steamboat smokestacks, pointing out that the chimneys of ocean vessels were of moderate height and still the furnaces did not suffocate for a draft. He waxed caustic about the wail over the too-wide spans: "If piers five hundred and twenty feet apart are too wide to serve as guides, there would be no means left for the bewildered navigator but to run it by the compass."[43]

When he came to the topic of the proposed canal he wrote in wrath. Canals happened to be an especially sore point with him at the moment, for his recent trip to the Mississippi delta had been to fight a proposed government plan to dig a canal near Fort St. Philip, below New Orleans, through a neck of land to an arm of the Gulf, forsaking, for the most part, the natural delta mouths because their channels over the Gulf bar needed deepening. And now another canal around the great stream! With malicious glee Eads jabbed at this, and at the morsel about how much better the board deemed it for a boat to jam into a pier than for its smokestack to hit an arch. He was having a very good time.

The board of engineers, although scathed by James Eads, pelted by letters from an irate public and snubbed by a chastened Secretary of War, were no whit discouraged. They held a second convention early in January, 1874, to consider estimates for their canal. They sat for four days while people on the two banks, awaiting the verdict, were fed by rumors dreadful or absurd: The Bridge was to be changed, mutilated. No, indeed, one newspaper editorial lent mock reassurance, the Bridge would be untouched, for he had invented a plan to lift the steamboats by balloons right over the arch tops and plop them down neatly on the other side.[44]

The board went solemnly on, extending their various plans. The canal, if dug, would be 1,400 feet long, not the original puny 800 feet. The drawspan would be wider than before, one end of it *to rest on the East Abutment!* The total cost—to be borne by the Bridge Company—would be $1,172,000. There were, the board admitted, alternatives that might be employed: to remove the west arch and substitute a truss or drawspan, or to remove all three arches and replace them with straight chord trusses—all this in reckless disregard that trusses would demand more piers to clutter the stream, and that drawspans had been outlawed. But, the report concluded with a magnificent sweep: "If it should prove that no change can be devised . . . then justice demands *the Bridge must come down!"*[45]

This supplementary report was hastened to General Humphreys, Chief of the Army Engineers, to accomplish what it might.

CHAPTER X

THE GREAT BRIDGE

WHILE the board of engineers sat in conclave to condemn his Bridge for a second time, James Eads had ridden through the winter streets to the levee and boarded the tender. Ice cakes thudded against the boat as it labored toward the workbarges. From its deck, through the snow and sleet driven by a furious wind, Eads watched the men moving about on the slippery arch runway, handling frozen ropes, brushing sleet from the coupling grooves. He had called this the Valley's Bridge, the world's, but today it belonged most of all to the men who crawled over its icy ways. Their feet like lead, their lungs stabbed by crystals of mist, they worked absorbedly. No one else took such a possessive interest in it. Yesterday Inspector Cooper had tripped on a plank, fallen ninety feet and vanished under the gray water. When he came up again he still clutched a pencil he had been about to use, and he begrudged the time it took to change his clothes before he could begin where he had left off.[1] This Bridge, built of stone and steel and the courage of men, surely could not be doomed by a board of diehards huddled about an open fire in a hotel parlor.

Yet, within a few days word sped through the cities on either bank of the river, and on into the hinterlands, that the board of engineers had decided that the Bridge would likely have to come down. This rumor lanced across a general chuckle over a spurious "opening" of the Bridge, an elaborate ten-dollars-a-plate banquet at the Laclede Hotel, featuring the beloved Swedish singer, Christine Nilsson, then performing in St. Louis. Invitations to the "opening" had utterly dumbfounded the Chief Engineer and

Bridge officials, none of whom recovered in time to attend. Only the charm and tact of the gracious singer carried the affair over the thin ice, while a curious public wondered what it was all about and a local newspaper informed them waggishly that it was likely a celebration of Jackson's victory over the British at the battle of New Orleans in 1814.[2] Now all smiles died away. Thousands of persons, perplexed and sullen, went to the levee and stood in the sharp wind to stare at the semiarches and the network of cables reaching from them to the temporary frame towers. All this work of years might be torn away.[3]

But nothing further was heard of the sensational wail of the board of engineers—it appears to have perished of neglect on the desk of the chastened Secretary of War. James Eads had given it scant attention. He was more concerned with the wedding of his youngest stepdaughter, small, blond, quiet Josephine. And a fine young man Josie was marrying, Wallace Estill McHenry, of old Maryland stock, part owner of the St. Louis *Daily Times,* and one of its editors. It was a home wedding, held at high noon on January 14, 1874.[4] Eads saw the young couple off on a train for Kentucky, the groom's birth state, and settled back to his work.

The superstructure was growing fast, the closing of the arches being accepted now as a simple task and extensible tubes as a matter of course. The Chief Engineer took time to make a valedictory address before the St. Louis Academy of Science, talking of seismographs, nebulae, molecules and atoms, of tides and winds, of a future when the air currents would be as completely charted as were the ocean currents, and aerial navigation "entirely practicable." And of the Mississippi River.[5] Then he left on a trip to New York to assure the eastern stockholders and bankers that the Bridge was nearly completed.

Captain Eads was away, and even the Bridge arches seemed to know it. He had hardly stretched out in his New York hotel bed when a telegram was brought to his door, a despairing message from Inspector Cooper—the arch ribs had begun to break, one crack after another had appeared. The crash of these words "was almost equal to a thunderbolt," Eads said later. After all the care he had taken, the Bridge was crumbling! He had virtually made every metal part himself, made it so strong that no load which could conceivably be laid upon it would overstrain it. Where was the flaw, the hidden weakness?

Pushing the turmoil from his mind, he put each fact through a merciless review, every metal part, every method of joining them, was marshaled sternly past—he was metal-hard himself and desperately strong as he made this ruthless inquisition. The verdict was unassailable: The Bridge was sound, the cause of breakage lay outside of it. The steel cables! Still attached to the arches—they had contracted in the severe cold and were pulling the ribs in a direction in which they were not intended to endure great strain.

Dashing off an order to Cooper to loosen all cables, James Eads took it to the telegraph office. Now he could sleep, the peril had been a myth. The next morning he met the investors confidently, assured them that the Bridge would soon be ready for use, and took the first train home.[6] He found the people on both shores excited, they had heard that the Bridge was breaking to bits. In spite of reassurances, they stamped and shivered on the levees in the following days to witness a disaster as the supporting cables began to vanish. They saw, instead, a miracle. The last cable was gone, the long lacy arches hung safely in mid-air, their tips charily poised on the stout foundations.

The great Bridge was nearly completed, but the final work lagged maddeningly. Appeals to Keystone to supply enough workmen pattered off the nonchalant contractors. Cynically James Eads suggested that Carnegie was angling for a further bonus. The St. Louis company clucked irritably over this, but arranged to meet Keystone officials in the East on February 5, 1874, the very day on which the first upright struts of the upper, or vehicular, roadway were put in place on the arch tops. At this meeting it was agreed that a bonus of a thousand dollars should be paid Keystone for *each day* prior to June 1 on which railroads and vehicles could use the Bridge. On the other hand, Keystone was to pay a penalty of five hundred dollars for each day after June 1 required to complete their contract. This proved a mighty stimulus to Andrew Carnegie. At once the arches swarmed with workers. In the crowding, several men fell into the river and one was drowned. The rat-tat of riveters never died from the frosty air. Struts and beams sprang to place. On April 15 the upper road was finished.

Andrew Carnegie, with his attorney and Manager Piper of Keystone, rushed to St. Louis upon an earnest mission: to collect part of the recently arranged bonus at once lest some delay thereafter dissolve it. The moment was perfect—James Eads was away. A long and clamorous parley with Bridge officials ended in an agreement by the St. Louisans to pay Keystone a pro rata bonus, provided that the upper roadway was thrown open to the public on the following day, April 18.[7]

It was late in the evening when the Bridge officials, worn out by Carnegie's insistence, had succumbed to these terms. Dr. Taussig, strained and tired, took what satisfaction he could in writing notices to the local newspapers that the upper roadway would be ready for use in the morning. This would be front-page banner-line news, hailed by everyone—it might even push the chronic hurray about the Siamese twins off the back pages. After he had delivered his notices and applied to the police department for a squad to handle the crowds at the Bridge approach, he dragged wearily home.

At midnight he was awakened and formally notified that Keystone had withdrawn their part of the agreement—they would not open the roadway, they had decided to hold all of the Bridge as security for any money due. Taussig was staggered. By now his notices would be in print, and no telling how many vehicles and pedestrians would jam the streets at the Bridge approach tomorrow. He could fancy the shouts of irate drivers, the rearing of horses, the traffic tangle. Early in the morning, with several other company officials, he started for the scene of trouble. The only encouraging note was a downpour of rain, the crowd would not be great. But when the little party reached the Bridge a new shock greeted them: Andrew Carnegie had had the workmen laboring since midnight tearing up the west end of the roadway, and now Manager Piper had drawn them up in battle array to resist any attempt to encroach upon the Bridge. Captain Eads had always made them feel that it was their Bridge, and this man called "Pipe" by his stubby, talkative little boss said that it was threatened.[8]

Suddenly the army of General Carnegie broke into excited babble and belligerent motions. They had spied the ordered police squad marching crisply toward the scene and thought that they were about to be attacked by an invading force. There was an excited outburst on each side, bawls and orders filled the air, a battle of sticks and stones was narrowly averted.

By good fortune the rain continued in torrents all day Sunday and Monday, so there were no disappointed crowds to turn away. Besides, it had become generally known that part of the roadway had been torn up and that the Bridge was under hostile guard. Public indignation over the "outrageous" action of Keystone mounted, but Andrew Carnegie remained quite untouched. On Tuesday he had the east end of the roadway pulled apart. Colonel Piper organized different watches, the men standing guard day and night. "We held the Bridge," Carnegie boasted years later. "Pipe made a splendid Horatius."[9]

James Eads, in Washington, was pelted with bulletins from his aides about the battle of the Bridge. Another time he might have enjoyed a good laugh over its absurdities, but now it was serious. There had been enough delay, bankers wanted their money, the people wanted their Bridge. He arranged a meeting in New York between representatives of the two companies. Andrew Carnegie left his armed forces under Piper's command and sped east. He peppered the meeting with words, but between volleys a compromise was reached: the upper roadway of the Bridge would be opened on Saturday, May 23.

Andrew Carnegie had come out well in this compromise. He was satisfied, and put his army to replacing what they had torn apart, and then to completing the roadway and sidewalks. On

Sunday, May 24, the sidewalks were opened to pedestrians. Singly, in couples, in groups, gentlemen in frock coats and high hats, ladies in Princess dresses, the flounce-bordered demitrains held up at one side, roustabouts and loafers, streamed over the Bridge. It was exhilarating, this treading on the very tops of the arches and towers, and looking away for miles up and down the sunlit river. The ferry boats shuttling from bank to bank already seemed relics of a past era. This stormy stretch of the Mississippi was bridged at last.[10]

On June 3 there was a formal opening of the Bridge's road for vehicles. At 5:30 in the morning, the four "spanking" bays of Joseph Gartside, hitched to the heaviest of coal wagons, stood at the entrance of the St. Louis side ready to be the first across. It was a proud moment for Mr. Gartside as he led the other waiting vehicles, while spectators, who had risen at dawn, cheered him.[11]

The long elevated approach on the east bank was ready for use, the nearly mile-long tunnel under the St. Louis streets had been about completed by James Andrews, who had taken over its construction when an earlier contractor abandoned it. The railroad tracks on the Bridge were being hurried to place. A few days later a pioneer locomotive crossed the Bridge carrying company officials, the Chief Engineer, and their guest, Chief of Staff General Sherman, who was in St. Louis arranging to move the United States Army Headquarters there because he suspected the integrity of Secretary of War Belknap in several instances—it had got so that Washington was not big enough to hold Sherman's blunt honesty and Belknap's wily intrigues. On the Illinois shore where a gang of laborers were laying the last stretch of rails, the visitors, done out in cutaway coats, stiff collars that punched under their jaws, and the inevitable tall hats, fell to with pick and mallet alongside the workmen, then stood back to let the General drive the last spike.[12]

St. Louis had held its breath for fear the Bridge would not be ready in time to celebrate on the Fourth of July. Preparations had been stewing for weeks, and hope pushed hope that nothing would go amiss. On July 1, James Eads had a private test of the Bridge made by sending a trainload of gravel and iron ore across it. His public test took place the following day, with crowds standing on the banks and upper roadway of the Bridge to watch it. The tale told by spectators when they returned home was sensational. Fourteen large locomotives, their tenders filled with coal and water, and all crowded with as many passengers as could cling to them, had crossed and recrossed the Bridge. First, seven had gone over, making a dramatic pause on each arch top. Then fourteen, seven on each track, had crossed side by side. Finally the fourteen had crossed in a single line. The shouts of the onlookers had been deafening.[13]

This test had been disappointing to James Eads, he had planned one far more convincing and spectacular. He had designed the Bridge to bear the weight "of the greatest number of people who could stand on the roadway above, and at the same time have each railroad track below covered from end to end with locomotives, and for this enormous load not to tax the ultimate strength of the Bridge more than one sixth of the strength of the steel of which the arches were constructed." But fourteen locomotives were all that he had been able to borrow.[14]

Fond of pageantry and parades as it had been since its early French days, St. Louis had never contrived a demonstration as elaborate as this Fourth of July celebration honoring the Bridge. Every kind of profession, trade, business and commodity would be represented by floats or costumed marchers—stove makers, saddlers, brewers, bakers, shirts, buggies, hemp bagging and harness, jewelry and lightning rods, shoes and soap, fraternal organizations, temperance clubs, German singing societies "six hundred strong," the United States cavalry from Jefferson Barracks, the National Guard, the fire departments. Bunting and flags were hung everywhere. A triumphal arch had been erected not far from the opening of the Bridge, and topping it, "tastefully decorated with evergreens and fifty feet high," was a medallion portrait of Captain Eads.[15]

News of all this had spread in every direction, and on July 3 people poured into town by wagon, horseback, train and boat. "Everybody was astonished at the outpouring of the thousands." The heat was suffocating. The visitors drifted to the Bridge, which was lighted by its lamps and a hazy moon. They wondered at its vastness, they recalled the many obstacles in the way of its building that had been surmounted by the determination of a courageous man. That night the hotels overflowed. Many slept in the lobbies and on the stairways, on trunks in the baggage rooms, on poolroom or saloon floors, in wagons, on the ground.

At dawn, Simpson's Battery moved to the riverfront and fired thirteen guns for the original thirteen states. At nine o'clock two of the guns were taken to the east side of the Bridge, where they fired alternately with the west side guns, a hundred resounding shots in all. Thereupon a fifteen-mile long procession started, conspicuous in it a full-rigged brig, the *James B. Eads*, drawn by eight horses and escorted by the customhouse officials. The long line passed under the triumphal arch and wound upward on the Bridge approach. Unfortunately, a fire alarm sounded and several engines dashed out of the line and off. Frightened horses reared, drivers yelled, women screamed. Dignity was restored and the parade went on, crossing the Bridge toward a triumphal arch and a colossal Statute of Liberty at the east end. The music of the bands, first one and then another, wafted back over the water to the tramp of many feet and the ring of ironshod hoofs. Songs rose and fell, banners and uniforms twinkled in the sun.[16]

James Eads had come to town early. From afar he could see the gigantic portrait on the triumphal arch. It grew higher and more terrifying as he neared it. That dome rising like an igloo toward the top of the frame was his own baldish head, the cataract splashing down from it was his beard! One of the mottoes read: "The Mississippi discovered by Marquette 1673; spanned by Captain Eads 1874."

There was a railroad excursion over the Bridge, a train of "palace cars" that delivered the celebrities of the occasion to the speakers' stand at the Celebration Pavilion on Washington Avenue at Third Street, President Grant, governors and mayors, legislators and financiers. There was much greeting and seating, the speakers facing a crowd that oozed into every street, bulged from every window and doorway. James Eads listened to the eulogies of himself and his work, remembering many things. When he rose to make his talk the storm of applause nearly abashed him. Flushed, he began:

"The love of praise is, I believe, common to all men, and whether it be a frailty or a virtue, I plead no exception from its fascination." He liked that opening sentence and was to use it on many another occasion when acclaim thundered at him. Then, subtly reminding his hearers of the lonely times when he and the Bridge were left nearly friendless, he recalled that when the first deep pier reached bedrock many persons had hastened to tell him how glad they were that his mind would be relieved.

"I felt no relief," he said, "for I *knew* that it must go there safely. When the first arch was closed, Mr. J. S. Morgan, of London . . . wrote me, hoping that the closing of the arch had made me as happy as it had him. I replied that the only happiness I felt was in the relief it afforded my friends, for I *knew* it would be all right. . . . Yesterday friends expressed to me their pleasure at the thought that my mind was relieved after testing the Bridge, but I felt no relief, because I had felt no anxiety."

He looked away at the Bridge, seeming to lose himself in it—he was still an almost inseparable part of it, he had drawn its plans, worked out the details, designed the building equipment, inventing twenty-three machines and devices for it, he had lobbied for it in Congress, pleaded with banks for loans, labored in a steel mill, prodded or coaxed contractors, he had been to the depths of the foundations, clung to nearly every part of the arches. "Yon graceful forms of stone and steel," Eads went on, "stand forth, not as the result of one man's talents, but as the crystallized thought of many, many minds, and as the enduring evidence of the toil of many hands." He spoke with regret of the men who had lost their lives in the construction of the

Bridge, fourteen of them from air pressure. He thanked all who had lent encouragement or money. Lusty cheers almost drowned the last words.[17]

When the speeches were over, the steamboats performed their part in the celebration, forming in lines facing the Bridge, the small craft in front—a curtsy called the "rainbow plan." In the evening elaborate fireworks were set off from the top of the Bridge, "a vast concourse" of people watching from the levee and upper decks of boats.[18]

Out of the shimmering illusion of national, if not world, dominance in which St. Louis had long wrapped itself, the Bridge emerged as "the one Great Reality." There it actually stood, beckoning the East into the new West. Of James Eads, the man who had turned the wistful dream of his fellows into a mighty structure, a thoughtful St. Louisan wrote: "He made speeches clad with imagery, but more deeply poetic was always his grandiose conception, and then his monumental execution."[19]

Now that applause of the Bridge rose on two continents, Messrs. Linville and Carnegie came forward to take the curtain calls. Gustav Lindenthal, who visited the Keystone Company after the Bridge was in use, wrote of the pride with which Mr. Linville spoke of the structure. Andrew Carnegie, assuming most of the credit for it, belittled James Eads in a magnanimous way: "A man may be possessed of great ability, and be a charming, interesting character, as Captain Eads undoubtedly was, and yet not be able to construct the first bridge of five-hundred foot spans over the Mississippi River, without availing himself of the scientific knowledge and practical experience of others."[20]

The St. Louis Bridge, acclaimed as the "greatest engineering feat of that sort" up to its time,[21] long continued to draw visitors from many countries. It was notable for its architectural beauty, for having the first pneumatic foundations in America and the deepest in the world, for being the first bridge anywhere to make an extensive use of steel, the first to use tubular chord members, and the first to depend, in building the superstructure, entirely upon the use of cantilever, with no false works employed.[22]

Twelve years after its completion, the Bridge was described in the Encyclopaedia Britannica as "one of the most remarkable structures in the world in character and magnitude." It had a profound influence upon bridge construction. In far countries piers rose above rushing streams where they had not dared before. In California, India, Europe, steel truss and arch spans were reminiscent of the patient months spent by a determined builder in a Philadelphia steel mill.

James Eads was well pleased with the Bridge as it stood, but he never quit thinking up possible improvements over his methods of building it, or of making them known. At times he appeared more alive to this than to the fact that, by reaching far ahead and overcoming difficulties never even challenged before, he had opened a new era in bridgebuilding.[23]

CHAPTER XI

THE GULF BAR

THE people of St. Louis never wearied of looking at their Bridge or of showing it off to visitors. They took their drives where they could gaze at the arches outlined sharply against sky and water at sunset, faintly in the starlight, or hanging ghostly above the mist that often rose over the stream and distant lowlands. On Sunday afternoons the upper roadway of the Bridge was a favorite resort, carriages and surreys rumbled along it, their occupants taking in the long reach of river on either hand, promenaders sauntered the footway. There was not another bridge like this in the world, St. Louisans knew—Eads Bridge, they called it. James Eads had raised it there in spite of racing currents, gorging ice and rabid opposition. They wondered idly or resentfully how long it would be before the railroads would end their boycott of it.

For a year after the grand opening celebration, the Bridge that was yet to play a heroic part in the development of the West did not rumble to a single locomotive. The railroads from the East had been accustomed to dumping freight and passengers on the east bank of the river, letting the cost of getting them across fall where it would, and they now had no idea of absorbing any of the Bridge toll, which was only about one-half the price that ferriage had long been. They would not adjust or compromise. As far as they were concerned, St. Louis was not on the railroad map, and they would continue to unload and reload when they reached the river, leaving the rest of the problem to trucks and ferries.[1]

Still the Bridge was marveled over by two continents, and its builder was one of the most discussed men in America. The *Scientific American* suggested him as a candidate for president of the United States, other journals and newspapers had joined in the movement.[2] But James Eads, who usually had a finger in politics, wanted no public office. He played the part of his town's most prominent citizen with simple artistry, doubtless enjoying the role and finding it useful. As a speaker on nearly every public occasion he could set forward a new project which he meant to launch, one that dwarfed the building of the Bridge. It lay nearer his heart, too, for it concerned the Mississippi as the valley's road to the sea.

The great trunk stream was more than a valley waterway, more than a national thoroughfare, he held, it was a world road reaching into waters that linked this country with Europe and Latin America, yet the door to it was half closed by the submerged cordon of mud in the Gulf outside of it. Large vessels could not enter it, the small ones that sifted through often had to endure ruinous delay. And nothing was done about it. Public distress over the choked Mississippi mouth was a century and a half old! As far back as 1726, men had tried to deepen the channel over the Gulf bar by raking its crest with harrows so that the Gulf littoral current, running parallel to the shore, might carry the loosened mud away. But the river, ever building the continent southward, had patched the bar with earth washed from its high banks by floods, and the harrowed channel was gone. A hundred and twenty-six years later, several Army engineers were sent to the delta to see what could be done about the barrier.

They found a strangely desolate scene. The river, which flowed deep and swift below Natchez to well below New Orleans, at last widened, slowed, and presently crawled in three branches, called passes, down the middle of three outstretched crooked fingers of reed-grown marsh and slimy beaches. Its brown waters fanned lazily into the blue Gulf, past enormous mud lumps, some of them acres in extent, that had erupted from its bed, and crept over the bar to let the last of their sediment slip from their lax grasp. Without enthusiasm the engineers had suggested that the bar outside of Southwest Pass, the largest branch, be attacked with harrows. If this proved ineffective, bucket dredging could be tried. Or false banks, jetties, might be built into the Gulf. In case all these failed, "a ship canal could be resorted to." Again harrowing was the method chosen and the natural 13-foot channel over the bar made five feet deeper. Three years after that, no trace of this new depth could be found.

Such a wail had gone up from valley shippers that Congress voted another sum to deepen this channel, and, if found desirable, the one leading from Pass à l'Outre. A mile-long single jetty, formed of a row of sheet piling, was built out from the east bank of Southwest Pass, the bar was harrowed and dredged, and a 21-foot channel secured. James Eads could imagine how the river had dawdled and laughed in the sun, stretched itself indolently away from the piling, spread its sediment with practiced skill and restored its shallow bed. Once more ships arriving in low-water season were closed out, departing ships closed in. Produce from upcountry gorged the New Orleans warehouses or rotted on the waterfront.

Another three years, and, in 1859, a party of New Orleans businessmen visited the delta and sent up a despairing cry: Thirty-five vessels were waiting inside the bar for high water to come and lift them out, seventeen waited outside to get in, and three ships sat grounded on the bar! This plaint was lost in the national din over abolition and secession. When Farragut's warships had to be dragged across the bar with their keels plowing it, the government had roused, for a moment, to the sad condition of the Mississippi's front doorstep, but the emergency passed and Washington forgot. The war ended, inland commerce reached out, ships from the whole Atlantic sought New Orleans. Up went a loud protest about the Gulf bar.[3]

It was while he was sparring with Mr. Boomer, James Eads recalled, that the government had set a steam dredge to work outside of Southwest Pass. The dredge created a depth of eighteen feet, but had to labor unendingly to hold it, so tireless was the river in its rebuilding. It was then that the valley began to talk of deserting the delta to cut a ship canal about seventy-five miles below New Orleans, from near Fort St. Philip eastward to Breton Sound, an arm of the Gulf. "A canal, a canal" had chimed on every hand. It would be so simple, all the delta worries cut off at one clean stroke! Two years later, in 1873, when he was prodding Keystone for Bridge parts, a three-day convention of United States congressmen, governors of states and others was held in St. Louis, beginning on May 13. On the second day of the convention James Eads, sent before it to offer a series of resolutions, presented a strong case for a better Mississippi channel. He presided at a banquet at the Southern Hotel on the third evening, and when the delegates left on an excursion, furnished by the railroads, to New Orleans and the delta, he was with them. They were bent upon a study of the canal plan, and James Eads was bent upon persuading them to discard that plan for one he had worked out.[4] A man-made ditch to replace the natural outlet of the great river system would never be dug if he could prevent it!

To many of the delegates the Mississippi had appeared a careless, notionate stream. James Eads, edging in an occasional story of his experiences on and under its surface, tried to show how relentless were its whims. This seemingly capricious stream followed a stern pattern, he said, established when the prehistoric torrent raged down from an inland sea in the north to tumble over a coral reef into an estuary that reached from the Gulf up through the continent as far, perhaps, as Cairo. To deal with this mighty river men had to take into consideration the log which it had kept through the ages. Part of this log was written in the shoaling beyond the delta passes. And there was the case of the so-called "Jump," a new outlet that had broken open about thirty years ago above the delta—when the break occurred, the water in it was a hundred feet deep, but by now shoaling had cut this depth down to four or five feet. Meanwhile it had formed over a hundred square miles of land upon which plantations flourished. Before the proposed canal could possibly be finished, the river would commence a subdelta formation at its mouth in Breton Sound.

There was no point in creating a new delta instead of improving the old one. The gentlemen found all this diverting—Captain Eads was always an interesting talker.

When he pointed over the Fort St. Philip region and stated with finality that the canal, with its necessary lock at the river end, and its masonry walls, could not be dug through the four and a half miles of marsh and two and a half miles of salt water in the short time nor for the thirteen million dollars that they estimated, they wondered at his vehemence. When he insisted, as the boat veered into Southwest Pass, that with a little aid from man in the form of proper jetties the Mississippi itself would dig an excellent channel through the Gulf bar at less cost than the proposed canal, they grew guarded. Or annoyed.

The pass, he had explained to them, in flaring from its quarter of a mile width to a mouth two miles wide, weakened its current until the suspended silt began to settle, the last of it sinking over the bar. In a year the Mississippi transported more than two hundred and seventy-six million cubic yards of material to its outlets, "equivalent to a prism one mile square and two hundred and sixty-eight feet high." The river's task was great—it was bearing the plains and hills, even parts of the Rockies from faraway Yellowstone, to fill in the sea. It kept a bed graven for itself through the land it built, and it could be coerced into etching this channel deeper at any point by constricting the stream with jetties. Struggling between narrowed banks that it could not escape, it would gouge furiously at its bottom, tossing the sediment ahead of it. If jetties extended the pass to the crest of the bar, the maddened current would cut a shipway faster, better and cheaper than a canal would provide.[5]

His listeners had gazed dubiously over the sluggish pass winding its way down its peninsula of mud. Mud lumps fifteen or twenty feet high stood in the Gulf, erupted from its soft bed. Nothing could look less promising, the very murmur of the reeds was depressing. True, the mouths of some European rivers had been successfully jettied, but none of those streams could be compared with the Mississippi, which was greater in volume than all the rivers of Europe combined, exclusive of the Volga. It was equal to three Ganges, nine Rhones, twenty-seven Seines, or eighty Tibers.[6] Yet Captain Eads talked of controlling it, of putting it to work, as casually as he might of turning on and off a faucet. Could this violent river be trusted to abet the jetties, even if the false banks did not sink of their own weight in the silt of the Gulf bottom? The bar itself, so soft that a man could thrust a cane deep into it, would swallow the jetty ends before they were completed. No, jetties were too experimental. Why risk public funds upon the unknown? A canal would be far less of a gamble.[7]

Their determination to abandon the river's own outlet was, to James Eads, unthinkable. He pleaded, reasoned, tried to explain away their suspicion of jetties. He had been overtired before he started on this trip, the days were hot and humid, the night wind chill. The old heaviness returned to sit on his chest, a cough kept interrupting him. The delegates waited stolidly for his impetuous arguments to finish. Almost no one was won over to his plan.

He had come back to St. Louis sorely disappointed. If he had not been so exhausted, he thought, if his lungs had not gone to bleeding, if the Bridge did not demand him so jealously, if his doctors had not shipped him off to sea, he might have done better by a deep natural outlet for the river. By the time he had returned home from Europe, in October of 1873, the valley had generally accepted the plan for a Fort St. Philip canal.[8]

James Eads felt that he could not endure having the river subjected to what he considered a mutilation, and, with the uncompleted Bridge still driving him, he began a singlehanded fight to get the canal scheme put aside for a jetty plan. By lectures and pamphlets he took the matter to the people whenever he could, describing in simple clarity the inexorable laws of flowing water and borne sediment, and outlining the control that man might wield over rivers, even the willful Mississippi. So lucid were his explanations that the most untutored men could understand them, yet experts over the world acclaimed them as "unsurpassed as engineering exposition."[9]

It was about at the time when the second conclave of Army

engineers and steamboatmen, in January of 1874, had seemed bent upon tearing down some or all of the bridge that James Eads had commenced to draw up a bill proposing to create by jetties a deep shipway through the Gulf bar outside of Southwest Pass. He could hardly find a free hour to give to this, but in February, hearing that an appropriation bill for eight million dollars to start the Fort St. Philip canal was coming up in Congress, he hurried to Washington.

There he offered to provide and to maintain for ten years a jetty channel 28 feet deep and 350 feet wide at the surface, from Southwest Pass out through the bar, for ten million dollars, only two million more than had been asked to *start* the canal, and five million less than the estimate for finishing the canal. He and his associates, Eads promised, would take the entire initial risk of financing his project, not a dollar to be paid by the government until he had secured a depth of 20 feet, whereupon he would receive a million dollars, and an added million for each additional 2 feet of depth until the 28 feet had been achieved. The remaining half of the ten millions was to be paid in ten annual installments on condition that the channel held its depth—a thing that the harrowed or dredged channels had never done.[10]

He was not surprised that opposition to his offer sprang from several quarters, or that the most violent attack upon it was made by Chief Humphreys of the Army Engineers. Humphreys, who had lost his tilts at the St. Louis Bridge, now bristled horrifically over the presumption of that man Eads who, always meddlesome, was pressing to take over the most important piece of engineering work ever considered by the government. Rashly, Humphreys vowed that the jetties would cost twenty-three million dollars. And one of his subordinates, Major Howell, in charge of dredging at Southwest Pass, and who was, moreover, the surveyor of the proposed canal, hastened to stir up New Orleans by prophesying that the jetties, if built, would certainly doom that port. They would increase the bar, he said, while at the same time pushing it rapidly seaward—in order to keep up with it, the jetties would have to be extended day upon day, year after year, and be built tirelessly upward as they sank ever deeper in the sand; the cost of the jetties would be colossal, their effect tragic.[11]

New Orleans was alarmed by this dark prophecy. Editorials in the local press accused James Eads of deliberately trying to choke their outlet to the sea in the interest of rail lines, "all his hopes and fortune being in the railroad Bridge at St. Louis." Businessmen of the city held excited meetings, made frantic speeches condemning the jetty plan. A group of them wrote to Eads begging him to drop his unsound scheme. Jetties, they warned, would sink in the Gulf bottom, currents would undermine them, sea worms devour them, and, with all that, they would push the bar ahead of them as they grew outward, closing New Orleans off from the world. The Chamber of Commerce sent two delegates, Professor C. G. Forshey, a scientist, and ex-Governor P. O. Hebert, to plead with Congress to save them from Eads and disaster. This southern port had not been so upset since General Benjamin F. Butler and his occupying troops had landed on the devastated waterfront.

At the Capital these two gentlemen talked to anyone who would listen, they distributed pamphlets, they addressed committees with sincere and desperate eloquence, pleading that the fate of their city be not risked in the hands of "an outsider," meaning, among other things, a civil, not Army, engineer. One of them appealed to a Senate group: "We have exhausted argument and laid before you the results of science and experience; we come now with prayer. Would you, can you, honorable Senators, at such a moment, contemplate or tolerate the half insane proposition of strangers who can know nothing of our inexorable enemy, to dam his waters at the mouth by jetties or wing dams that must inevitably send back the flood waters like a tide to the very city of New Orleans, or beyond, and complete the impending destruction? . . . Do not, we pray, permit us to be destroyed, and that without remedy."

As a final thrust at the jetty dragon the two earnest crusaders distributed through the House of Representatives printed handbills that blazoned in scarehead capitals: "MUD LUMP BLOCK-

ADE—Forty-seven vessels blockaded at Southwest Pass, and one hoisted upon a Mud Lump that has suddenly reared its head right across the channel. . . ."[12]

Exasperated by all this well-meant fervor, James Eads began to lay about him. There was just one thing that could madden him to acrimonious bitterness: opposition to any measure that would better the Mississippi or its commerce.[13] In a somewhat tardy reply to the letter of the New Orleans businessmen he explained calmly the laws of river hydraulics, then abruptly changed his tone to one of jabbing ridicule:

"One of the most distinguished agents of your Chamber of Commerce here, Professor Forshey, tells the Congress of the U. S. that he has been taught modesty and humility in the presence of the gigantic torrent, and gives evidence of his modesty by claiming, in his pamphlet, that he has devoted much more time and personal labor to the investigation of the physics and phenomena of the river than 'any other one man *living or dead.*' While your ex-Governor Hebert, with no less modesty, has informed committees that he graduated at West Point at the head of his class in which were the living Sherman and the dead Thomas, and that he thinks it would be very unwise to trust so great a work as this to the crude notions of 'an outsider.'

"Had the Governor not graduated at the head of his class and learned at a tender age all that is worth knowing in this world, he might still acquire some knowledge by returning to West Point; for he would now be taught facts and principles in his profession there drawn from the experience of this *outsider,* and would have them explained to him by the writings, diagrams and models that were supplied by him at the particular request of that institution. While this insider [Hebert] was displaying precocity by beating Sherman and Thomas at school, and the indefatigable Professor was experimenting with a keg having two valves in it to find out what was going on at the bottom of the river, and was publishing to the world . . . that sand was not carried in suspension, I was daily learning the falsity of these theories by hard work on the river bottom, in the diving bell."

No doubt of it, James Eads, irritated as he felt, was enjoying himself. He proceeded: "Your people seem to have forgotten that the barque *Mary Ellen* was wrecked on the bar of Southwest Pass, and that after fruitless efforts to recover her cargo by others, it was by Eads' and Nelson's labors twenty years ago that the property was saved. The practical knowledge acquired in that work, occupying sixty to ninety days, enables me to speak with more certainty of the ability of the bar to sustain my proposed works than could be given by any theories founded on assumed conditions which do not exist."

He made light of the claim that the jetties would sink into the bar because a man could push a pole into it—locomotive trains carried ponderous loads over just such ground. Then, his confidence mounting over his temper: "Whether or not in my dealings with men I possess as much modesty as the eminent representatives of your Chamber do, I am sure I have not learned 'modesty and humility in the presence of the gigantic torrent.' Nor do I believe that it can be controlled by modesty and humility. . . . I should have a low estimate of the mental powers conferred by the Creator upon man if I did not believe him capable of curbing, controlling and directing the Mississippi, according to his pleasure, from the Gulf of Mexico to the most attenuated rivulet. . . .

"Your distinguished Professor, for whom I really entertain great respect, says to the Committee, in his pamphlet, that he 'came before men of judgment to confess his disasters and failures.' We differ. I have no confession of disaster or failure to make, for in my dealings with the Mississippi I have had none. The learned Professor candidly declares that 'whenever the deep abrasions, beyond the human eye, turned toward his works, he has learned to retire before their dread forces and place his feeble works beyond the reach of their devourings.'"

Eads began to tower now: "What I know of the Mississippi is facts, and facts are the uncut jewels which grind false theories to powder. . . . Your representatives here have now incontinently retired behind their earthworks—the *mud lumps*. From this sole remaining stronghold they have just issued the proclamation I herewith enclose, headed 'Mud Lump Blockade!' I will leave them there for the present. . . . When reflecting on the interest which paid agents have in prolonging this controversy, regardless of how the nation will suffer, I am tempted to quote the suggestive scriptural exclamation, 'Doth the wild ass bray when he hath grass, or loweth the ox over his fodder?'"

Storming on to a charge that he was in league with the railroads that had paid for the St. Louis Bridge: "Not one dollar was ever supplied in aid of its construction by any railroad. . . . If I am permitted to proceed, I shall certainly open the mouth of your river, and will double the value of every foot of ground in your city."[14]

These two hostile camps, James Eads on one side and the city of New Orleans on the other, were eventually to clasp hands in a common interest—the welfare of the great river's commerce. But now Eads's virile campaign for his plan and the understandable fear which Major Howells had aroused in a community dwelling on a low strip of land nearly islanded by water and saved from flood only by the strictest vigilance whipped up a whirlwind in Washington that seemed likely to blow off the tall hats and flop the dignified coattails of many a legislator. And it would be a long time before the gust would finally die down.

On the day that the jetty bill came up before the House of Representatives, a letter written by General Humphreys was circulated among the members. Apparently forgetting that he had declared the jetties would cost twenty-three million dollars to build, he now insisted that James Eads, in charging ten millions for their construction and ten years of maintenance, would put six or seven million dollars in his own pockets. Besides this, the bar, he wailed, which already advanced at the rate of hundreds of feet a year, would advance ten times faster than ever before and would have to be pursued full tilt by the jetties. This letter, plus the hubbub kept up by the two New Orleans gentlemen against the "outsider," drowned out a committee report favoring the jetties. The canal bill passed in the House.[15]

Its elated backers rushed it to the Senate, where the jetty bill also was to come up. James Eads fought inch by inch for his bill, throwing every ounce of his strength into it. His old friend, Congressman Erastus Wells, and Senator Carl Schurz, of Missouri, stood beside him in his efforts. Schurz said that it would certainly be no calamity for this channel matter to be placed in the hands of an outsider—in European countries that were highly military, works of a nonmilitary character were planned and put through by civil engineers, and surely the Army engineers had had a full fling at this river problem. "For thirty-seven years," he recalled, "they have been planning and reporting, and scratching and scraping at the mouth of the Mississippi, and today the depth of water is no greater."[16]

So forceful was James Eads's campaign that a Senate committee asked to be discharged from further consideration of the Fort St. Philip Canal. But Eads foresaw that there would be a deadlock over the two bills, so he hastily drew up a third to substitute for them. This bill provided for a commission of seven engineers, both military and civil, to study the delta problem and recommend a solution. The measure was pushed through.

For six months the appointed commission pondered the matter, only to admit finally that they were unable to cope with the questions raised "in this most difficult branch of engineering." They wanted to inspect various improved rivers in Europe before making a decision. James Eads put aside his writing and lecturing in behalf of the jetties and sailed for Europe, too, independently of the commission, taking with him James Andrews, who had been the contractor for the Bridge foundations and the St. Louis tunnel. He visited jetties at the mouths of the rivers Oder, Wipper, Persante, Vistula, Danube, Pregel and Rhone. He relished stalking down facts at any time, and this hunt was especially to his liking, for what he found bore out his arguments. Returning home in high spirits, he found that the press had largely swung around to his side and was setting forth the very views he had carried to the people at every opportunity for several years.[17]

When he went to Washington in December, 1874, for the opening of Congress, the chances for his Southwest Pass Jetty plan looked golden—it had leaked out broadly that the commission of engineers, back from Europe, had been converted to it. Eads was elated over the prospect that the Mississippi would have its shipway to the Gulf, a wide, deep channel that would serve commerce for a century at least.

He was roughly jolted out of this complacency by the commission's report. To be sure, they recommended his jetty design, endorsing all that he had claimed for it; but they indicated that the improvement should be made, not at big Southwest Pass, for which all his specifications had been drawn, but at narrow shallow South Pass, only one-fourth as large and doubly obstructed. The bar outside of South Pass was only eight feet underwater, and there was a serious shoal at the head of the pass, the removal of which, disturbing the whole torrent of the Mississippi, would involve the most hazardous piece of hydraulic engineering ever undertaken anywhere. James Eads did not see how he could fit his design to this little pass, or make the same guarantees of depth and width as he had put forth. Above all, he was disturbed over the fact that a channel at narrow South Pass could not be increased in width and depth when the size of ships had outgrown it.[18]

He beat back his disappointment and resolved that the large pass, not this totally inadequate one, would be opened. He made a new proposal to Congress to jetty the mouth of large Southwest Pass, going to extremes to make it attractive: He would create a channel thirty feet deep (not twenty-eight) over the Gulf bar for eight million dollars, and maintain it for twenty years (not ten) for one hundred and fifty thousand dollars a year. When he set his revised offer before Congress he was confident that his rash generosity would not be declined.

It was not declined. Eagerly Congress accepted the program for a deeper channel, longer maintenance and reduced payments, but stipulated that the jetties should be built at the little doubly choked South Pass.[19]

Shocked by this callous substitution, James Eads appealed to the commission of engineers, about to dissolve, to hear him. He talked before them for two and a half hours, protesting that he had never said he could produce a channel thirty feet deep over the bar outside the small pass, or that he would undertake the perilous work at its head. He could not, he said, offer the same guarantees that he had for the work at the large pass, nor attempt it for the same price. Furthermore, he was loath to spend effort and funds on an outlet that commerce must outgrow in a generation. He was very tired after he had finished this plea, but hopeful. All common sense and logic were on his side.

Up before Congress the whole matter came again to be hashed over. A decision was reached barely before the session adjourned. Awaiting the outcome eagerly, Eads was stunned to learn that the small pass was still stipulated, but that all guarantees he had offered in the case of the large pass must hold good. Moreover, the work must be done for three million dollars less than he had proposed, and the channel maintained for thirty thousand dollars less annually than even the commission had estimated it could be done for.[20]

Faced by the choice of undertaking the nearly impossible or of abandoning all hope of opening the Mississippi mouth to ocean shipping, Eads, after a surging conflict within himself, accepted the sorry terms left him. He would build the jetties at South Pass.

To the valley only one much-blazoned fact stood out—the jetty bill had gone through. It had not seemed possible that any man could fight past the powerful opposition that had ranged up against the plan. Captain Eads had done that. He had cleared the river of sunken wrecks, he had bridged one of the wildest stretches of it, and now he was going to open its way to the sea. The great stream had not had such a champion since Henry Shreve had cleaned the snags and the Fulton monopoly out of it. Men called the news from boat to boat: the river was going to have a deep channel through the Gulf bar. Farmers, insurance brokers and banks discussed it. On March 23, 1875, three weeks after the passage of the Jetty Act, four hundred prominent men from up and down the middle valley gathered in St. Louis for a banquet at the Southern Hotel to pay tribute to the man who had fought the river's fight. Pomp, ceremony and ringing speeches were the order of the moment.[21]

James Eads was surpised at the demonstration. He felt almost that he had failed the river when he did not get the large pass specified for the jetty building. The poor choice of site had been made in the name of economy, he explained when he rose to reply to the ovation, adding wryly: "If it be found that the greater pass shall ultimately require the same treatment, it will prove a false economy."

With his customary frankness, he took issue with a tone of skepticism that lurked beneath the enthusiasm of valley folk. In going about the organization of a South Pass Jetty Company he had encountered far more hope than confidence. On all sides he had been reminded that every attempt to create a permanent channel through the Gulf bar had ended in failure. Men who hoped without faith would, he intimated, furnish feeble support to this tremendous enterprise.

Gathering fire, he struck out at all embedded doubt, seeming to grow in stature as his voice rang out: "If the profession of an engineer were not based upon exact science, I might tremble for the result. But every atom that moves onward in the river, from the moment it leaves its home amid crystal springs or mountainsnows, throughout the fifteen hundred leagues of its devious pathways, until it is finally lost in the vast waters of the Gulf, is controlled by laws as fixed and certain as those which direct the heavenly bodies. . . . I therefore undertake the work based upon the constant ordinances of God Himself; and so certain as he will spare my life and faculties for two years more, I will give the Mississippi River . . . a deep, open, safe and permanent outlet to the sea."[22]

For the moment his hearers believed him.

CHAPTER XII

WILLOW WALLS

It was May, of 1875, summer in the delta country, when James Eads set out from New Orleans to the Gulf, James Andrews with him. The sky was hot and shimmering, the river flickered in the sun or nudged the dark reflections of cypress trees that waded out from the low banks to dangle a veil of long gray moss over the water. After a while the stream widened and two long points of willow-fringed land stretched far up into it, dividing it into three main branches. The boat turned into South Pass, which reached shallow arms around a shoal and wound on, sluggish and glassy, between flat banks on which reeds reared ten to twelve feet high. Driftwood piled against the banks, alligators dozed on the rotting logs, herons stood about on the spongy mud beyond the reed border, pelicans flew heavily back and forth or perched on distant mud lumps, curious gulls wheeled overhead. The pass broadened, yawning at last with its two-mile wide mouth, the last borders of the reed-grown strip nearly lost in a wilderness of water. Nothing rose anywhere above the flat monotony but a few fishermen's huts and a lighthouse.[1]

To James Eads the scene had a strange primeval beauty. This was the river's own domain, it had stretched these fingers of land into the Gulf, it could overflow every foot of them, spreading silt over them, widening them here, pinching them there—at places the Gulf inlets were only three hundred feet from the pass. The level of the Gulf coast, which was lowering two feet in a century along most of its ragged line from the Rio Grande to Florida, was sinking here at the more rapid rate of two feet in thirty-eight years.[2] But the river kept it built up almost to high-water stage, cutting its shallow bed through it and continuing between walls of salt water well into the Gulf. This was a sorry outlet compared with the greater passes, but into it the largest ships of the Atlantic would soon sail, and the rivers above it pulse with quickened commerce.

A small schooner was heading from the Gulf toward the pass, the *Research,* with a Coast Survey detail recently sent here to make soundings. James Eads welcomed the captain aboard his boat and looked over the charts handed to him. They bore out his conviction that he had undertaken a herculean but not impossible task. On the way back to St. Louis he pored over charts and mapped his campaign against the barrier, listing the equipment that James Andrews was to gather or have constructed.

In early June they had a fleet of tugs, barges, little steamboats, steam launches, yawls and two immense floating pile drivers ready. The motley flotilla, some of it spiny with derricks and small pile drivers, some groaning under loads of machinery, lumber or coal, pulled out of New Orleans at midnight of the 12th, snaked down South Pass before dawn, silencing the bullfrog chorus that rose mightily from the marsh, and anchored at the east side of the mouth where only a few yards of land divided the river from the Gulf inlet.

Men swarmed the boats and mud beach as they began to build a work settlement that Andrews named Port Eads. Hammers clanged on the blacksmith boat, axes rang on lumber, tugs whistled signals, engines chugged. One by one the pile drivers were put to work, their thuds scattering the herons from distant mud flats. Piles to which the boats would be moored, piles to serve as foundations for a landing wharf, for a warehouse, boardinghouse, office and cottages had to be driven down. The sun shone hot, steam rose from the marsh. The men sank in mud, fought off mosquitoes that hung about them in gray clouds at dusk, ate their supper on the temporary mess barge and crawled into their bunks.

Five days later the floating pile drivers began to lay two lines of piles to border the jetties. These walls would stand about a thousand feet apart, their width, at the top, increasing from ten feet near shore to fifty feet at the sea end. The east jetty would be about two and a third miles long, the west one somewhat shorter, the channel curving slightly, since the west natural bank of the pass reached farther into the Gulf. It was not many days until all was ready for the real work on the jetties.[3]

In building the jetties to constrict and deepen the channel, he was but following, James Eads had repeatedly said, the laws which the river itself demonstrated. And he planned to use, for the most part, the materials that the river often employed in building banks and islands. He and the river would carry on the construction together, he furnishing the two walls of piled willow-brush mattresses, the stream filling their every crevice with sediment. When they had been firmly packed with mud he would have them flanked with rubblestone and capped with concrete blocks. The cramped stream would dig at its bed, the quickened current would bear the loosened sediment well over the bar into deep water.

Willows were few along South Pass, but they grew in thickets along the "Jump," above the delta, and there the willow cutters were stationed, eating and sleeping aboard boats. All day they waded about, sinking in the marsh as they cut the brush which they slung in bundles over their shoulders and carried to a waiting barge. The bargemen heaped it on the broad deck a dozen feet high, the butts along the middle, the tips hanging over the side. A tug moved the barge down to Port Eads and sidled it to the ways on which the mattresses were made.

Meanwhile the mattress frames had been formed of yellow pine strips brought from nearby stacks in small trucks. The strips were laid in nine or ten parallel rows five feet apart, lapping ends spiked together to a length of a hundred feet. Stakes driven into holes in the strips stood up two feet. Now the willows were passed by the bargemen to the gangs on the ways and laid across the strips between the stakes, the brushy tops overhanging the frames several feet. When the mattress was piled about eighteen inches thick, another mattress was laid crosswise upon it, binding strips with holes in them were forced down over the stakes and secured by wedges.

The double mattress, frowsy with protruding brush, was hitched to a tug and pulled, floating, to position inside the jetty piling. A barge of rubble was brought alongside and the stone spread over a strip of the mattress; when this edge sank beneath the surface, the barge moved over it and rubble was laid on another strip. After the whole mattress had sunk somewhat, the mooring cables were let go and the weighted brush went to the bottom, against the piles. The next mattress was laid to the sea end of this one, and so on out. Other mattresses would be heaped upon these, building the wall well abovewater.

James Eads, who liked to see his plans work out, doubtless enjoyed watching the mattresses made, the brush for them moving along the ways, the finished ones towed off to place. Brush mattresses, or fascines, had long been used in Europe, but with the "Dutch system" employed there it took two days to put together a mattress as large as these. By the process which he and James Andrews had invented, it took only two hours. Barges of brush were always waiting at one end of the ways, a tug was always towing a mattress off the other end. The work was going fast.

But the elements were to lay a heavy hand on it now and then. Torrential summer rains bogged down the willow cutting, it was hard to get men to wade in the sucking mud. The mattressways devoured the occasional bargeload of brush, and waited. A severe September gale, whistling out of the Caribbean, wrecked more than a hundred mattresses floating in position along the east jetty, waiting to be sunk. November fogs closed in. The world at Port Eads was a place of sounds hanging in a gray mist—a tugboat whistle, the clack of hammers, the scraping of willows, the honking of wild geese flying southward, and the scant talk of men. Sometimes there boomed from the gray the foghorn of a ship making for Southwest Pass where Major Howell and his crew maintained an eighteen-foot channel through the bar by constant dredging. The willow cutting and mattress making went on in slow monotony. Another severe storm, on December 2, 1875, destroyed more of the work. James Eads had expected these hazards, they were included in the risk that he, and not the government, bore.

But he was content. For despite backsets, the uncompressed walls of brush and silt, growing beneath the brown water that spread into the Gulf, had created a lively current which had scoured out nearly two million cubic yards of mud from between them and carried it to sea. And by late February, 1876, when the work was only eight months old, the once eight-foot channel over South Pass bar had deepened to thirteen feet![4]

It disturbed him, though, that a dull pessimism about the jetties had been instilled into the valley. General Humphreys, never relenting, had written four essays, termed by him "memorandums," to prove that the jetties would end in worse than failure, creating a greater bar in the Gulf, closing the Mississippi off from the world. He had fitted the memorandums into his official report to Congress, waited until they had been duly printed as part of a Congressional document, then excerpted them, illustrated them with maps, bound them in pamphlets and distributed them over the country. This was followed by a series of anonymous statements in newspapers that a new shoal was forming outside of South Pass. New Orleans, lately somewhat reconciled to the jetties, was again hostile to them, St. Louis skeptical, the smaller ports uneasy.

The General's campaign would have to be defeated, James Eads realized, and he thought it might be done more effectively without words. The passage of a good-sized ship through the jetty channel would be more eloquent than any protest he could make. He arranged with the skipper of the *Mattie Atwood,* bound from New Orleans with a cargo of cotton for Revel, Russia, to set to sea through the jetties, assuring him that there would be a depth of thirteen feet over the bar at low tide. He pinned a good deal of hope on this scheme. But when the *Mattie Atwood* arrived at the pass on the appointed day, she was loaded to draw thirteen *and a half* feet. The tide was out. The ship grounded on the bar and had to sit there until the tide came in the following day.[5]

Such a boon as this mishap was to General Humphreys and his aides! Their clamor over it swelled until some of the most loyal advocates of the jetty project were shaken and refused to put further money into it. The company's finances were in a precarious state. James Eads tried to show that the channel improvement was phenomenal, but was met everywhere by cold doubt. He suggested that anyone interested enough to criticize might well come to South Pass and see for himself. The St. Louis stockholders decided to do this, but in order to make it a friendly investigation, they got up an excursion, chartering the *Grand Republic* (the recently rebuilt and refurnished *Great Republic*), now the handsomest floating palace on the river, and invited the families of investors and members of the press.

On an April day, 1876, the passengers trooped aboard the luxurious boat. The orchestra played, handkerchiefs waved, a crowd on the levee cheered as the steamer, sparkling with new paint, backed out from the wharves. The river was chalky and boisterous under a pale sky, the willows along the Illinois banks were greening, the forest on the high Missouri shore was lacy with buds and fragile new leaves. There was promenading on the upper decks, leisurely formal meals in the long saloon cabin, dancing in the starlight, love-making in the shadows, gambling in the card room. And there were earnest conjectures by a handful of bankers and shippers on the chances of a deep channel at the river mouth.

On into the south the steamboat and its load of excursionists glided. Long green islands slashed the river, canebrakes reared their sun-sifted green canopies on the low shores, live oaks waved the intruders away with gray arms of moss from the swamp secrets they guarded, the scent of magnolia and honeysuckle rode the breeze, woodpeckers kept up a rat-tat inland. The boat stopped at New Orleans, took on more passengers and, on April 26, set on downstream for the delta.

At the mouth of South Pass James Eads awaited it, almost boyishly eager to show off the jetty channel, for a recent scouring by the quickened current had increased the depth to sixteen feet. He was proud, too, of the clean, comfortable little settlement, with its white cottages and flower-bordered plank walk along the river. His assistant, Elmer Corthell, and most of Port Eads watched with him. Someone spied a sign of motion far up the pass, but instead of the lofty chimneys and broad white bow of the steamboat emerging from the shimmer of sun and mist, a small boat moved swiftly toward them, raced on by into the Gulf, darted about in a zigzag course, stopped here and there, and came back to moor nearby. This whimsical craft was Major Howell's steam launch, sent the thirty miles from the dredging base at Southwest Pass to add an unexpected touch to the steamboat excursion. In its cabin Howell's assistant, Captain Collins, was making a hasty chart.

While James Eads and his party puzzled over the launch, the *Grand Republic* was sighted. It came on majestically, looming enormous against the flatlands, its chimneys growing taller above the reeds. When it stopped at the landing it dwarfed the port. Eads and his associates boarded it and the big steamboat moved on through the jetty channel, over the bar, tossed on the Gulf a short distance and turned back. The passengers were cheering gaily as it moored, hailing the new deep opening to the sea. They surrounded James Eads with congratulations.

On the edge of this gathering appeared Captain Collins, armed with his rapidly drawn chart which he managed to show to those nearest him. It positively refuted the claim that Eads had just made of a sixteen-foot channel, it threw doubt upon even a thirteen-foot depth. Worse still, it reported a serious shoaling beyond the bar, ostensibly caused by the jetties. A murmur of consternation set up about Collins, the circle around the jetty builder began to melt away at the edges and form into whispering groups. The boasts of Captain Eads were false! The stockholders were being deceived, the newspapers hoaxed!

When James Eads realized what was going on, he was dumfounded. He produced the latest report made by his engineers, but these outsiders could not give a chart the stamp of officialdom as could Captain Collins of the Army engineers. The civilian depth figures were unconvincing and failed to allay the suspicions that had been raised. Helplessly Eads saw the excursion, which he had hoped would build new confidence in his plan, destroy what faith there had been. The affair ended in coldness and distrust. When the *Grand Republic* reached New Orleans on its return trip, some of the passengers threw their Jetty Company stock on the market at half its face value.[6]

A sense of injustice, even of hounding misfortune, pulled at James Eads. There was the *Mattie Atwood* incident, and now this cheap trickery of the false chart. It would take all his ingenuity to keep the Jetty Company from falling apart. He wrote to the New Orleans *Times* insisting that the channel depth was exactly as he had represented it, and denying any new shoal ahead of the bar. Instantly a statement appeared in another newspaper that the bar had simply removed from one place to another. Eads replied that this was a "malicious falsehood . . . founded in ignorance and malice." He went on bitterly: "I was led to believe that every intelligent citizen of New Orleans wanted deep water at the mouth of the river. . . . We have been at work less than eleven months, spending our own money . . . and have obtained a *permanent* increase in the depth of South Pass bar. . . . And yet one would suppose from the fault-finding that we were increasing the expense of the government, or wasting the money of all these grumblers." He enclosed a telegram just received from his aides—the depth over the bar had reached *seventeen feet*.

The fracas was not over. Major Howell came back at "the recent insidious attacks of Mr. Eads." His newspaper letter grew spicier as it proceeded: "In his usual brow-beating manner, with which the people of New Orleans are so familiar, he [Eads] attempts to choke off investigation. . . . There was no volunteered attempt . . . to influence the minds of the St. Louis delegates against their 'Josh,' his works or his teachings. . . . On the day of the *Grand Republic* splurge there was at South Pass only a channel of twelve feet. . . . The nucleus of a new bar existed one thousand feet in front of the jetties."[7]

Eddies of this controversy widened until they reached Captain Gager, master of the *Hudson,* a large coastwise vessel of the Cromwell Line. Irked with the chronic difficulties at the Mississippi mouth, he had taken a marked interest in Eads's jetty plan. Too often he had approached Southwest Pass to find a fleet of sailing ships and steamers lined up and waiting at anchor to see the great dredge move out of the channel so that they might, each in turn, be hauled through it by "the tugging harpies who were always waiting for their prey." After a storm they could not even be dragged across the new sediment that filled the ditch, they had to wait until the dredge had cleared a channel. Captain Gager sent word to Eads that he would bring the *Hudson* in by South Pass on its next trip to New Orleans.

This was his chance, James Eads believed, to furnish the proof he needed. On May 12, the day scheduled for the coming of the big boat he kept a constant watch for it. The tide rolled in, and no steamer with it. The tide turned outward. There would be channel enough at low tide, but the drive of the outgoing surge would pull a heavy adverse current through the jetties, setting up a real difficulty. At last, a speck in the sunlight, the *Hudson* appeared from the southeast. Eads went with the coast pilot in a small boat out beyond the bar to meet it and go aboard. So much depended upon this entry that the skipper must be warned of the swift current against him.

Captain Gager frowned down the information. "Head her for the jetties!" he ordered, and rang for full speed ahead.[8]

The ship pushed in, the outrushing tide breaking against the prow in a white spray, and rode triumphantly to the landing, then on to New Orleans. That port was astounded. A large ship entering the river mouth without being tugged over shoals! And through little South Pass! The news was peddled about. Sentiment began to roll up behind the jetties, mounting when other large vessels, upon hearing that the *Hudson* had entered the Mississippi without paying tribute to the towing "harpies," made for this new entrance.

The momentary public excitement over the deepened channel lapped across the nation, enveloping a royal visitor, Dom Pedro II, the fifty-one-year-old Emperor of Brazil, who had come to the United States to attend the Centennial Exposition at Philadelphia, and to learn all he could of American progress. After a whirlwind, but thorough, trip to Chicago, on to California and back east, he journeyed to St. Louis, about the middle of May, 1876, with his wife, the homely but much-beloved Doña Theresa, and his modest suite. He whizzed through the sights of the city and countryside, and boarded the *Grand Republic* for a trip to New Orleans and the jetties.[9]

The day after the royal arrival in New Orleans, James Eads was conducted to the Emperor's apartment in the St. Charles Hotel by the Brazilian Consul. He was not surprised to find Dom Pedro, who was a Prince of Braganza and Bourbon, fair, tall and portly, his gray hair and beard framing a serious but pleasant face. The newspapers had made much of the Emperor's appearance, as well as of his plain dress and the democratic manner in which he was touring this democratic land. Dom Pedro repeated his desire to look over the jetties, and Eads invited him and his suite to make an excursion there on his small yacht *Julia*. The time for the trip was fixed for Saturday, two days hence.[10]

All the next day and the next, the white horses of the Emperor's carriage literally galloped from place to place in and about the city, and on Saturday evening kicked up the dust of the levee as they made for the yacht at the foot of Julia Street. There were only a dozen or so in the excursion party, including several New Orleans stockholders in the Jetty Company, for the quarters were cramped. James Eads and his royal guest were on excellent terms at once. They had much in common, both were omnivorous readers, inveterate early risers, impatient of the flitting of time, brimming with energy, dignified, determined and simple. Still there were moments when the grave Dom Pedro must have puzzled over this profound American engineer who joked so easily and laughed so lightly. As the *Julia* puffed through the night down the silvered stream, with the cry of night birds, twitter of insects and croaking of frogs dying ahead of it and rising behind it, Dom Pedro, his black slouch hat pulled over his wide forehead, told the jetty builder of the wretched condition of some of the Brazilian river mouths, with their growing harbor bars.[11]

The yacht reached Port Eads at daylight, and in the gray cool of dawn James Eads began showing the party about the workbarges and mattressways. They sailed on out through the jetties to the bar, sounding here and there. Dom Pedro was amazed at the depth secured in so short a time and at such comparatively small expense. He had soon seen everything and was ready to go back to New Orleans and be off to Mobile. On the way upriver he requested James Eads to come to Brazil and improve some of the river mouths there. Eads expressed his appreciation of the offer, but declined it. He was too occupied with his efforts here to undertake anything so far away. Dom Pedro was politely insistent—simple as was his manner, there were moments when he could draw himself up in regal pride as though enveloped in his ceremonial feather robe of state. He finally accepted Eads's recommendation of W. Milnor Roberts (who had been substitute chief engineer for the bridge) for his projects.[12]

Dom Pedro's outspoken astonishment at the success of the hardly begun jetties, and the fact that ships were weekly seeking them out, raised a passing wave of public approval of them. But rumors of their failure only rose louder and thicker: there might be a ship channel today, they mouthed, but it would have vanished tomorrow and conditions would be worse than ever, far worse, and all this government money wasted! The valley was being too easily duped, it was listening to the most visionary spendthrift it had ever produced!

To James Eads's dismay, this propaganda sponged out the flattering facts as if they did not exist. Grimly setting about to have an official acknowledgment placed on his claims, he requested the Coast Survey to have their Captain Marinden, then taking soundings in the Gulf, extend his inquiries over the bar and through the jetties. This was refused by General Comstock, who now strode masterfully upon the scene. Comstock, stationed at Detroit, oddly had authority over surveys at the Gulf, and he was General Humphreys's own man. Eads held his chagrin, wondering which way to move next. It happened that Comstock's assistant, Captain Brown, was presently sounding at the South Pass bar. Hopefully Eads asked for a chart of his findings. Brown replied that he would have to report first to General Comstock. Then, hearing one day in New Orleans that Comstock had just gone through that city on his way to the jetties, Eads hurried after him, met him leaving South Pass and asked him for an official report with which to meet the stories that were threatening to ruin the jetty project. Comstock said he could give his information only to General Humphreys, in Washington.

Choking back his exasperation, Eads begged Comstock to go back with him to Port Eads while he telegraphed Secretary of War Alphonso Taft to instruct them to sound the jetty channel and the Gulf beyond it together, a matter of only a few hours' work. Grudgingly Comstock returned to Port Eads, making it clear that he would give up no more than the remainder of the day to this nonsense. That would be enough, Eads believed.

He sent a telegram to Secretary Taft, explaining the vital need of checking the rumors that were rolling up to swamp the jetty program. By evening, when no reply came, Comstock left. Four days later a dispatch came from Taft saying that General Comstock was at the jetties to survey the channel, had doubtless done it, and "a copy of his findings would be forwarded as soon as received."[13]

As soon as received! Eads blazed. After the usual roundabout journey of such reports! For the results of surveys, "carefully kept secret from Mr. Eads and his assistants . . . were transmitted by Captain Brown to General Comstock, at Detroit; by him to the Chief of Engineers at Washington; by him to the Secretary of War; by him to Congress; by Congress to the public printer; and by him back to Congress. They were then so old as to be of little interest or value to anyone." As soon as received—after a sixty-day jaunt, while the valley buzzed with damning misrepresentation and the last of the lean funds for the jetties ran out!

Still bent upon getting some saving evidence, James Eads telegraphed the Superintendent of Coast Survey, in Washington, to have Captain Marinden, again surveying beyond the jetties, furnish him a chart, since he, Eads, had provided the little steamer that Marinden was using on the express condition that he could have the result—a bargain made with General Comstock. The answer came back: "Regret Marinden cannot furnish results. General Comstock will give all information required by law." As was expected, Comstock refused to give it.

Infuriated by the wall of secrecy that had been set up across his path, James Eads looked about to see how he could batter it down. His pleas, arguments and explanations of his predicament had, so far, been futile. Chief Humphreys sat in mighty disapproval of the jetties, Comstock played up to him, and their subordinates could only dodge inquiries and not commit themselves. Eads decided upon a flank attack. He addressed an open letter to Secretary of War Taft, pointing out that his inability to get a single official report with which to parry the propaganda against the jetties was endangering the financing of the tremendous work he had undertaken.

"We are assuming all risks and expending our own money under a grant which gives us no power to deceive the government even if we desired to," he wrote. "We were entitled to eight months to commence the work and thirty months within which to secure twenty feet of depth, yet before fifteen months have elapsed the largest coasting steamers trading to New Orleans have been sent to sea over the bar on which scarcely eight feet of water could be found last year."

Sternly he demanded that future reports from the Coast Survey concerning the jetties be made directly to the Secretary of War, and not to General Humphreys, who had assumed a control expressly forbidden by the Jetty Act in a clause inserted especially to hold Humphreys at bay. And he insisted that a comparative chart of the soundings that had been made in and about the pass in the last year be furnished him. Secretary of War Taft seems to have ignored this letter, but he was shortly succeeded in office by J. D. Cameron, who at once sought to bar Humphreys and his faithful Major Howells from further interference with the jetty work, and requested that the Coast Survey provide Captain Eads the needed charts.[14]

This should have been the end of the episode, but was not. The Superintendent of Coast Survey, "wishing to avoid conflict with the U. S. Engineers," declared that he would not supply the charts even to the War Department. The baffled Secretary of War then referred the exasperated jetty builder to the Secretary of the Treasury, who promptly referred him back to the Secretary of War.

While this game of dodge was going on, the channel over the South Pass bar reached a depth of nineteen feet, a foot deeper than the dredged shipway at Southwest Pass. But James Eads knew that it would be a dreary time before he could get this new depth certified to the public—he had not yet been able to confirm the sixteen-foot channel of May. And the croaking chorus against his plan was rising higher . . . the jetties were a failure, worse than a failure . . . they were building an impassible barrier beyond them . . . imprisoning the Mississippi, bringing ruin . . . ruin. His creditors beset him for interest payments he could not make, his few supporters dwindled. He winced at the resentment that rose from every boat landing against the meddler who was converting the Mississippi into an inland lake. Desperately he turned to the House of Representatives, outlining "the excessive injustice" done his project.

In contrast to the stony coldness met elsewhere, the response here was warm and vigorous. A resolution, unanimously passed, directed the Secretary of the Treasury to furnish the much-discussed chart.[15]

When James Eads spread the chart on the desk in his New Orleans office he forgot the August heat that hung motionless in the room, he no longer heard the clatter of mule cars or rumble of wagons in the street. This square of paper, with its few waving lines and stale figures, would disprove the damaging tales about the jetties that had been released like pigeons from a coop, for it shed the magic of official aura on the very figures he had claimed. It showed that exactly where his detractors had located the alleged new bar the now lively current had scoured out sixty thousand cubic yards of silt bottom, leaving the Gulf there deeper than ever.[16]

Sparing no names or feelings, official or otherwise, James Eads laid about him with the chart. One by one the adverse tales about the jetties fluttered back home to roost. Public tension over the river outlet relaxed. In New Orleans, some who had opposed Eads with all their might rallied tentatively to his support, others who were still wary of his daring project felt themselves drawn to this man who fought them so relentlessly for what he believed was the river's good but met them with disarming friendliness on all other points. However antagonistic they might feel when he was afar, they found themselves irresistibly attracted to him when he was at hand.

Dr. Souchon, a New Orleans physician, admitted this with an anecdote which he liked to tell afterwards. Earlier in this year, 1876, the doctor, returning by sea, "despondent, misanthropic," from a health trip to the East, had placed his chair in a secluded spot on the rear deck of the steamer. On the second day out, a pretty young girl seated herself not far from him and was soon joined by a man more than twice her age, but evidently, from his manner, not her father. The way in which he read to her, giggled over the book, chatted and joked, "pressing his suit," and the girl's evident gay enjoyment of it, irritated Souchon. For two days more he fumed over the recurring scene, then the "old flirt" vanished. Just as the doctor was hoping that he had seen the last of him, a steward came up and asked him to attend an ill passenger. Armed with his medicine kit, he was ushered into the stateroom of the patient—the "old flirt" himself. Souchon ministered ungraciously at first, but was "gradually and insensibly won over." Back on deck, he learned that the "modest, unassuming man" was Captain Eads, chaperoning the daughter of one of his jetty contractors on her way to visit her father at South Pass. When Eads was well enough to lie on deck, Dr. Souchon found himself edging eagerly into the growing court that gathered about his cot.[17]

Timidly New Orleans hospitality reached out to the man who was sincere, at least, in his belief that he was bettering the river outlet. And James Eads, who made it a practice "never to refuse a hand held out to him in friendship," no matter if it had pounded a table in denunciation of him an hour before, was pleased over the invitations that sprinkled his business mail. He ate at burdened tables, sat on balconies and looked away through the scrubby palms at the river, while banjos tinkled distantly and the swaying chant of Negroes rose and fell. He listened to proud fragments of Creole history, he told his best anecdotes, leading the laughs heartily, he quoted Robert Burns in his best broad Scotch "with exquisite grace," set out his newest poetic find, Sir Edwin Arnold, reciting long passages from his *Light of Asia,* the life story of Buddha. And he pictured the day when the largest ships in the world would enter South Pass in a steady stream to moor yonder at the waterfront. A fascinating man, New Orleans appraised him, even if visionary and overoptimistic. No one else could make an evening so glamorous, many a hostess explained as she got off an invitation to the jetty builder.[18]

But Captain Eads was never long in New Orleans. The work

at South Pass demanded him consumingly. The crying need of this scant outlet was water, more water. A deep channel would be of doubtful service unless there was water enough at all times to fill it. Even after he had closed Grand Bayou, which had been diverting and wasting one-fourth of the waters of the pass, he had to contrive a way to entice a greater flow from the river. To this end, he began contracting the mouth of Pass à l'Outre by six hundred feet; and in order to make the most of the water thus gained, he was closing, by means of a dam, one side of the narrow channel around the shoal at South Pass head. While the work was being done, however, much of the water in South Pass had to be deflected into the other passes. All this made up a nearly crushing effort—it was hard for James Eads not to resent being put to it by the obstinate selection of this inadequate pass for a deep shipway. Now, while the outsider strove to surmount the enormous difficulties that had so lightly been set up, the government engineers looked on from afar, wondering curiously if it really could be done.

The most knowing civil engineers shuddered for the outcome as Eads swung the giant torrent of the river aside, tinkered at the shoal, and swung it back again. This manipulation of the mighty stream required far more courage than the building of the jetties at the mouth. W. Milnor Roberts characterized it as "the most difficult piece of engineering in river hydraulics which the world has ever seen." And Elmer Corthell, assistant engineer at Port Eads, wrote of it: "The river and its entire volume, and its great width of a mile and three quarters, thirty feet deep, with a strong current, was bridled by cheap, rough mattresses laid as sills upon the beds of the great Passes, guiding and holding the volume of water into South Pass as required, deepening the bar at its head. . . . There is no instance, indeed, in the world where such a vast volume of water is placed under such absolute control by the engineer through methods so economic and simple."[19]

But while this work was going on, ships had to be shut out. The jeering sidelines made the most of this: the jetty experiment had played the deuce in South Pass . . . ships were being warned away from there . . . the mess might prove hopeless, the pass ruined! A new wave of pessimism surged up the rivers. James Eads had to turn from his exacting task to combat it. In August, of 1876, he reported:

"In seventeen months after the passage of the Act, and within fourteen months of the commencement of the work, the jetties have solved the problem presented at the mouth of the river. . . . They have not been overturned by mud-lumps, nor swallowed up in quicksands, nor undermined by the river current. . . . Over three million cubic yards of earth have been swept away from between them into the Gulf."[20]

Six weeks later, on October 5, 1876, a big day dawned on South Pass—the channel through the jetties and over the bar was *twenty feet deep.* A twenty-foot channel, and a payment due of one half of the million dollars owed by the government! The other half of the million was to be held back until the project had proved its promised worth over the years. James Eads sent his bill to Washington, and waited expectantly.[21]

And waited in perplexity. No check came. General Comstock, at Detroit, had set his foot down on any payment until the channel at the head of the pass was as deep as that over the bar. This stung Eads to fury. His contract called only for a deep channel over the Gulf bar, but the work at the pass head needed doing and was being done, gratis, a tremendous undertaking, hampered by lack of credit. Comstock would never know what a struggle had been made at the shoaled head—in the midst of it, Resident Engineer G. W. R. Bayley had died of a heart attack, a probable sacrifice. In a fiery protest to the War Department, Eads laid out the facts, stressing that the money was urgently needed for strengthening the sea ends of the jetties with rubblestone before the winter storm waves attacked them. It should have been done before this, he emphasized, but he had not had funds for it.

Generals Comstock and Humphreys sat immovable, blocking the payment. No money was due the Jetty Company, they repeated stonily, overriding Secretary of War Cameron's reminder that they had nothing to do with the jetties. The whole matter was put up to President Grant. Bombarded with the conflicting claims, the President had a commission of three Army engineers sent to inspect the work. Five weeks after the payment had fallen due, this commission arrived in New Orleans. They went over maps and papers with Eads in his office there, and journeyed on to the pass. When they made their report they admitted that Captain Eads had no obligation whatever at the head of the pass, but that he was, nevertheless, creating a ship channel there. James Eads read this part of the report with satisfaction, only to stiffen with hurt as the concluding paragraphs grew severe about the lack of stone bolstering at the sea end of the jetties. As though there had been any money for this! Or was still any for it! Was not that the prime reason that he wanted the payment hurried? Always sensitive, despite his driving will, Eads took this chiding to heart as a personal humiliation.[22]

While the legal niceties of whether or not any money was really due the Jetty Company were being picked over in Washington, time drifted along. In January, James Eads went to the Capital to take a hand in the matter. Three months had been wasted, and not a penny paid for the gigantic work and its admitted phenomenal result! By January 19, 1877, the Secretary of War, after much prodding, got around to reminding the Speaker of the House that a payment had been due Captain Eads since the October before, and that no appropriation of money had as yet been made for it. On the following day Eads himself sent a demand that, since no money had been made ready for him, he be paid in bonds, his contract stipulating that this could be done. Several days later the Secretary of the Treasury announced that he would not pay in bonds and thereby increase the interest-bearing debt of the nation while there was plenty of money in the treasury. A deadlock, and the jetties still destitute! James Eads went to the Secretary of the Treasury with a personal plea, met a cool refusal, lost his temper and said some very tart things for which he afterwards apologized.[23] The Secretary held his ground manfully.

Days passed. And weeks. Eads felt caught in a vise. Time, the inexorable, pushed at him. The unfinished dam and unbolstered jetty ends could wash away while officials hemmed and hawed: Oh, yes, Captain Eads sitting outside for an audience—he was always in some antechamber . . . the payment on the jetties, to be sure, it would be looked into in good time. Eads and his jetty payment—would the man never let up?

On February 6, a committee at last reported that, since Congress had failed to make any appropriation, Captain Eads was entitled to be paid in bonds.[24] About the middle of the month the bonds were actually handed over. Eads hurried south to see for himself how much his works had suffered, for ocean storms and river floods had not waited with him. A late December gale had demolished the unsupported jetty wing dams in the Gulf and lifted off some of the top courses of mattresses, the river had gone wild and spilled over the unfinished dam at the pass head, doing what mischief it could. Every day that the payment had been delayed had added to the enormous bulk of work that would have to be done again.[25]

Driving himself mercilessly, James Eads collected material and began to repair the damage. He bought what he dared, then had to stand aside and see the hovering debts, like vultures, devour the rest of the bonds. The jetty project settled again into poverty and toil.

CHAPTER XIII

THE PROUD LITTLE PASS

T HE stormy seas subsided, the gray fog that had hidden the world from Port Eads vanished. It was early March of 1877. The jetty builder, prowling the village restlessly, took in the glint of sunlight on the white cottages, the green strip of shrubs along the plank walk bordering the river, the wet beach yellow against the bright blue of the Gulf. Wild ducks rode the waves with sea gulls or tailed in long lines northward, white fleets of pelicans swam the bay. Red-winged blackbirds, too ravenous after a Caribbean flight to be frightened by the noise of engines, swarmed the beaches and reed jungles, their musical calls mingling with the shouts of the mattress makers. Men towed finished mattresses to place, hoists lifted them upon the jetties, which now reached abovewater. Men sidled stone-laden flatboats along the sea ends and threw rubble about the walls to reinforce them, men were constructing temporary wing dams within the walls to slow the sides of the river current so that more silt would be dropped in the willows. A pilot boat flitted out to the Gulf to usher a ship in—many ocean vessels sailed through South Pass now. The valley, which had "a preponderance of the population and business of the whole United States,"[1] would have a splendid outlet to the sea if money could be found with which to carry on the work. But where could funds be raised, with the bankers and investors tuned only to the adverse hubbub that General Humphreys and his submissive aides kept alive?

James Eads went to New Orleans for another attempt to raise a loan there. He went from bank to bank, arguing, pleading.

He waited days for conferences to be held, he heard the final refusals. At last, in April, 1877, he was promised two hundred thousand dollars, but before the loan could be closed a wail was published in a local newspaper that the much-flaunted twenty-foot jetty channel was rapidly dwindling. At once the loan was "postponed."[2] To meet his payroll Eads had to beg petty sums where he could, renewing the notes each month. But a substantial fund was needed to carry on the work at the head of the pass, he had had the waters there deflected and he could not keep them so while he waited indefinitely for material he could not buy. Summer had slipped in and was passing. Money must be had somehow from somewhere.

Without optimism James Eads started to St. Louis to lay his nearly desperate situation before bankers there. Damaging press dispatches winged after him, arriving in time to do their blithe bit in defeating him. Anyway, it was not a propitious moment to make appeals in St. Louis, for a nation-wide railroad strike had halted business there and filled the city with unrest. A radical element called the International, or Commune, had grasped this occasion to launch a campaign of violence. Gathering in noisy groups they denounced "all regular organized forces," which they contended were "against the future of the human race."[3] Their flag, a red handkerchief aloft on a fishing pole, in the forefront, they marched to the levee four hundred strong and went by ferry—snubbing the capitalist Bridge—to East St. Louis to lend their sympathy and oratory to the strikers there. Given a frigid reception by the orderly railroad men, whose only use for a red flag, they said, was to stop a train, the zealous disbelievers in private property recrossed the river to St. Louis, attracted a number of Negro roustabouts to their mob, and set off on an orgy of pillage and loot, entering homes and stores, terrorizing the populace. Troops had been called out on July 27, and the Communist rebellion quashed, but no one knew when it might burst forth again.[4] The banks were more than nervous. They would not lend money now on anything, least of all on the much-condemned jetties.

There was nowhere left to turn. Letting disappointment roll over him, James Eads saw his dream of a ship channel from the Mississippi to the sea fading away. He began a telegram to Elmer Corthell, now resident engineer at Port Eads: "Discharge the whole force—" It was hard to go on. No one, it seemed, cared enough about a free outlet for the river to make a sacrifice for it—unless, perhaps, the jetty laborers themselves, as loyal a company of men as had ever undertaken a great work. For a moment Eads lost himself in this, his respect for workingmen amounted nearly to reverence. "Toil is the foe of vice," he had once said to a group of students, "and I am sometimes tempted to believe that it is more powerful in preventing wickedness than religion itself. Honesty is the favorite companion of labor."[5]

He hastily finished the message, but not in the way he had first intended. "Discharge the whole force, except those necessary to protect property," it read, "unless they are willing to work for certificates payable on receipt of the twenty-two-foot channel payment."[6]

On a sultry August (1877) afternoon at Port Eads the Resident Engineer sent word to all the jetty workers to report to him at once. Tired and hot, their faces flushed, sweat-drenched shirts clinging to their shoulders, the men filed into the office. The sun beat down on the wooden building as they shuffled to make room for each other along the walls. The very silence was suffocating as Corthell, at his desk, paused to steady his voice before reading the telegram from his chief. The first words came as a blast to the men, echoes of them mingled with the roar of the waves against the mattress walls. Discharge the whole force, stop building the jetties—just as the work was beginning to count big, the channel growing deeper each day, the water pushing so fast between the walls that it was slashing a deep ditch through the bar!

"Unless they are willing to work for certificates payable on receipt . . ." the voice went on. Obstinate faces relaxed, men turned to each other. Why not? They were housed and fed—if they deserted the jetties they would never feel right again. The Captain had told them more than once how much they were doing for the river, for their country. They would go on, and wait for their pay.

All but two of the seventy-six men lined up to sign the pact. Elmer Corthell, moved by the scene, wrote afterwards: "Little did the great Valley realize that its vast commercial interests were being advanced by a few laboring men at the mouth of the Mississippi River, to whom great credit should be given for their devotion to duty and faith in the ultimate success of the jetties."[7]

The men worked with spurred intent, the restless current tore at its bed. All was well, confidence had become a habit. One could count the few weeks until the 22-foot depth would be reached. But on a day in mid-September the sky greened ominously, gulls flew landward in clouds, a Caribbean gale swept across the Gulf, buffeted the workboats, shook the houses and ripped viciously at the jetty tops. It raged for three days, while the workers watched anxiously as waves twelve to fifteen feet high flung against the piling-flanked brush walls to tear from them their newly piled rubble. Before the wind had died, barges were shuttling back and forth, the storm damage was being repaired and new work hurried along.

James Eads's lips grew straight as he pondered the unpaid men working absorbedly, jealous of every inch of progress and worried over a stubborn area in the jetty bed near the sea end where the scour was too slow. To correct this bulge in the channel floor and hasten the longed-for payday, he put a dipper dredge to work there, but the sea was stormy and the dredge could not hold its place against the waves. The disagreeable fact had to be faced that there was no dredge in existence which could do the work in that fury of water, and the autumn seas growing heavier each day. Delay of the government payment would discourage the men, and it would pile up interest. There was nothing for him to do, Eads saw, but invent a dredge boat for the emergency, and he would have to contrive it quickly, for the fate of the jetties raced against time.[8]

In the last week of October, 1877, James Eads went to Cincinnati to meet the big iron dredge which he had designed in feverish haste and had constructed at Pittsburgh on a rush order, at a cost of a hundred and fifty thousand dollars. Its builder, Mr. W. D. Carroll, was bringing it down the Ohio. The members of the Cincinnati "Exchange" asked Eads to address its members on the subject of the jetties. He accepted the invitation gladly, for he liked talking about the jetties. But when he mounted the platform he found himself a target for heckling questions and derisive cries kept up noisily by a group of dissenters throughout his address, making it nearly impossible for others to hear him. "Wharf rats would have been more courteous," a local newspaper commented in disgust. Eads had received a telegram that day which he read proudly: the jetty channel had reached a depth of twenty-one and a half feet.[9] But the words were lost in the uproar.

Although South Pass awaited him, James Eads appears to have stopped in St. Louis on his way. His younger daughter, small, brown-haired, blue-eyed Martha, "Little Mattie," was marrying young Edward Montague Switzer, the son of one of his friends, on October 30. It was a disappointment to the St. Louis social set that it was a quiet wedding, and it was probably a disappointment to James Eads, who liked to do the utmost for his daughters. But with hard times gripping the country, with his own pockets empty, with the jetty men working for promises, this was no time for social splendor. Nothing but the lights streaming from every window of the Compton Hill mansion advertised the event to the world.[10]

On November 22, only two months after the dipper dredge had been defeated by the waves, James Eads saw his large sea dredge, the *G. W. R. Bayley,* put to work. It was entirely different in principle from the government dredge boats. Employing a powerful centrifugal pump and a large suction pipe, it made the water the vehicle for raising the desposit from the bed. The pump could raise one hundred and thirty tons of water a minute, twelve feet high. The four tanks in the body of the boat that received the deposit, which was carried out to deep water and discharged, each ended with a hopper in the bottom, the closing valve being manipulated by a hydraulic jack, supplied by a hydraulic ram. Except for the centrifugal pump, the other outstanding features of the dredge were invented by Eads. Three weeks after this dredge was installed, by the middle of December, the stubborn places in the jetty channel had yielded, there was a 22-foot depth from South Pass out over the Gulf bar.[11]

The long strain was over, a second payment due. James Eads got off his statement to Washington—surely there would be no delay of this installment. He was amazed when a grave official debate arose over the question of whether anything was due him or the men who had been giving their time and labor on trust. The Jetty Act specified that the channel be achieved by jetties and auxiliary works—was the new dredge a proper auxiliary? Leisurely a commission was sent to the pass to investigate.

After a thorough exploration the commission reported that, while the dredge was an auxiliary appliance and not an auxiliary work, such as a dam, its effect was only one per cent of the whole, "utterly insignificant" compared with that of the jetties. And the 22-foot channel claimed by Captain Eads actually existed. Another while, and the payment was made. Eads gave the men their back wages and parceled out all but a fraction of the remainder to hungry debts.[12]

The story that a once miserable eight-foot channel at one of the Mississippi passes had been nearly tripled in depth refuted the rumors of failure that had been kept flying over the country. Visitors began flocking to the jetties as they had to the Bridge. The walls of willow brush and river mud, as yet barely above low tide, sounded flimsy and looked insignificant, but they had given the valley such an outlet as it had never had before. During the past year, 1877, five hundred and eighty-seven ocean vessels had gone through South Pass with no waiting at the bar. The number would increase with every foot of depth.[13]

One of the visitors early in 1878, Dr. J. S. Baldwin, of Jacksonville, Florida, had come to Port Eads on an urgent mission. A bar outside the mouth of beautiful St. Johns River closed out all but the smallest ships from the good estuary harbor of Jacksonville, twenty-four miles from the sea, Baldwin told Eads, and he had been trying for a quarter of a century to arouse public interest in a better channel over that bar. Upon hearing that large ships were coming and going daily through lately choked South Pass, he had come here to persuade the engineer who had performed this wonder to do the same for the St. Johns. Eads, striving against many odds to carry on his Mississippi project, was not minded to listen to anyone's troubles about any other stream, but there was no escaping Dr. Baldwin. He was mild but insistent, he kept on and on: If Captain Eads would only glance over the St. Johns delta and bar—the river's welfare was important not only to Florida but to much of the South. Reluctantly Eads consented to analyze the difficulties of the St. Johns but not to apply the remedy. Dr. Baldwin, glad of even this much, hurried home to raise a thousand dollars by popular subscription to offer for the service.[14]

It was March when James Eads arrived in Jacksonville. He was met by Dr. Baldwin and some of his associates, driven about the countryside through upland pine and lowland oak woods, then launched upon his survey. He found the St. Johns a silent river, its waters moving gravely, with no semblance to the Mississippi's bluster. Palmettos shrank back from the water, stretching their many-fingered hands toward each other, the Spanish bayonet bristled its spears in straggling companies. Live oaks ventured near the banks, their gnarled limbs waving streamers of moss at the cypress that waded in the river edge as if about to invade the endless border of lily pads. Cows standing breast-deep in the water munched a coarse grass that grew beneath the ripples "like a flooded meadow," dragonflies darted about in the sunlight. Across the wide stream the other shore seemed a faintly golden mist against the sky.[15]

After he had studied and sounded the river, following it to where it widened into the sea, James Eads showed his companions that this bar was of sea sands driven back and forth by currents, rather than of river silt. He gave his opinion that two converging jetties—made of palmetto brush, since no willows were at hand —the walls rising above the water and reaching from the mainland to the bar, would give a channel twenty feet deep at average flood tide. This would accommodate the ships that were likely to seek that port. The cost would be one and three-quarters millions.[16]

Again Eads insisted that he would not have time to undertake the work himself. But Jacksonville, grateful for his advice and encouragement, presented him with the thousand dollars that Dr. Baldwin had raised for him, wined and dined him, named a street for him, and were loath to let him go.[17]

While he was drifting down picturesque St. Johns or riding through the misty Florida woods, while he warmed to attentions and got off some excellent stories, James Eads had been torn in his soul over the Mississippi jetties. For a year a conviction had haunted him that the severe terms of his contract were squeezing the life out of his project. In his zeal to provide the river with a deep outlet he had accepted whatever heavy conditions were laid upon him, and had undertaken the impossible when he pledged to furnish at South Pass the 350-foot wide channel he had planned for Southwest Pass. Perhaps he had believed that he could surmount the towering difficulties—certainly he had not foreseen that the animosity of a few men would strip him of credit, or that government payments in time of peace would be so ruinously slow.

In May of 1878, well aware that he was poking a hornet's nest, he wrote Secretary of War McCrary, asking that some of the Jetty Act conditions be eased, pointing out that he had been crowded into shouldering the task of creating in the little pass a channel designed for the large one. He was convinced now, he said, that a channel three hundred feet wide was all that South Pass could safely sustain. Its slope was greater than that of the other passes, and any exaggerated enlargement of it might draw so great a volume of water from the river as to endanger the jetties. As for the finances, he had completed eighty per cent of the entire work and received but twenty per cent of the money he had laid out, although customarily the government paid for ninety per cent of any work as it progressed. He boldly asked that certain deferred payments amounting to a million and a quarter dollars be granted him, which would still leave a million held back by the government for security. He went on:

"We have changed the little Pass into a grand channel of commerce through which the largest shipping that visits New Orleans floats in safety. We have striven to carry through by private means and individual hazard the largest and most important public work ever undertaken by the government and have foreborne under many discouragements to ask for relief."[18]

Out buzzed the hornets, the indefatigable General Humphreys in the lead humming that the jetties had been a miserable failure from first to last. Civil engineers were in the swarm, too. One of these, Henry F. Knapp, of New York, who had already made speeches and printed a pamphlet denouncing the work at South Pass, wrote to the New York *Daily Tribune:* "As an hydraulics engineer, Eads is simply an idiot. . . . I will assure and comprehensively prove that jetties are humbugs. . . . If it will make it any more impressive, I will say that I am ready to put a few thousand dollars on any and all of the above assertions." But an editorial in the same paper held that, considering Captain Eads had been "compelled to offer the government a bargain as one-sided as a jug handle," his requests were reasonable.[19]

Believing that the fate of the jetties might hang upon the response to his plea, James Eads waited tensely. The decision that finally reached him eased his financial strain, allowing him five hundred thousand dollars to apply on his debts and an equal amount with which to carry on the work, but it left him still suspended in doubt about his requested change in the channel width—a board of engineers would be sent to the pass in a few months, when the current epidemic of yellow fever in the South had died down, to look into all the jetty factors and report upon them.

The hum of denunciation of the "South Pass fiasco" rose more shrilly when it leaked out that James Eads had been granted some financial relief: the country was being hoodwinked, the jetties would pull the national treasury down with them into quicksand; their perpetrator, oily of tongue and fluid of pen, had wheedled another fortune out of Congress and was after still more, don't forget that! Sensitive to this carping, Eads jabbed back briefly in the press, and dismissed it. He was engrossed in designing the equipment for topping the brush walls with concrete blocks.

Slightly as he had been able to lift the financial pressure, new life flowed in the jetty works. Tugs puffed back and forth towing flats of rubble to be thrown against the willow walls, barges brought sand, cement, gravel and lumber. The project that had seemed to most of the country so haphazard and forlorn would soon be completed. But the men were now restless. Strictly quarantined against yellow fever, no mail or newspapers reached them, and their food had grown monotonous because the stores were low. Every ship, towboat and barge would be closed out until the plague abated.

The August sun, steaming the mist that hovered over river and gulf, seemed harder to bear than in other years, and the threat of hunger at last played havoc with morale. A boat had to be sent to New Orleans for supplies. Every precaution was taken, but yellow fever was a stowaway on the return trip. When the first man at Port Eads went down with it, the others were permitted to leave if they wanted to. Of the hundred who remained, half were stricken and eleven died. James Eads, so adamant in his determinations, was softhearted about the men who carried through his plans—some of the plague victims had stood by the jetties when there was no money to pay them. He was wrung by their suffering. The death of William Nelson, commander of the sea dredge, builder of the bridge caissons and one-time "wrecking" partner, seized by the malady on his way to St. Louis, shook him. It was hard to go on contriving equipment for making the jetty tops with death stalking all around him.[20]

As the convalescent men grew strong enough to work, and the scattered ones returned, they proceeded with the rubble flanking of the jetties. It looked to them as though they might have a lifelong job at this, so greedily did the water swallow one bargeload of stone after another. But at last it was gorged, and the making of the concrete blocks could take the fore. These blocks had to be enormous, for the storm waves of the past winter had made a game of tossing off the temporary covering stones that weighed up to three thousand pounds apiece. The concrete mixer, steam elevator and dumping cars were lined up, strings of barges were pulling in with sand, cement and crushed rock that would make up the concrete, and great open boxes were built on top of the jetties in beds prepared for them. Now the mixer was clattering and the cars dumping the heavy mush into boxes that devoured it almost as hungrily as the water had the rubblestone. This crowning wall was to be 4 to 12 feet wide and 5 to 7 feet high, some of the blocks at the sea end weighing 260 tons apiece![21]

In the midst of this work, with the concrete mixer grinding, the elevator screeching, stone rattling and dumping cars grunting like mad elephants, autumn, of 1878, brought the long-awaited board of Army engineers to investigate the need of further financial aid and channel dimension changes. James Eads welcomed them sincerely. His want of funds was critical, the machinery and material for the concrete blocks piling up new obligations. He showed them by sketches and figures that he had now completed *ninety* per cent of the work and should be paid (aside from the million dollars to be held back by the government for security) $1,773,000, a sum that still would not lift from him all the debts he had incurred. The board visited all parts of the work, expressed themselves as elated over the excellence they found, and left.[22]

For the first time since he had begun the enormous task, serenity enveloped the jetty builder. His plan was fully vindicated, the heavy hand of official suspicion and disapproval was lifted. There were days in which the river seemed to bound joyously through its new channel, the Gulf waves danced past the willow walls, the reeds murmured happily to each other. Help would come from Washington, the crushing debts would be cleared away.

When a copy of the board's report reached James Eads it staggered him. The document was astounding. Enthusiastically it praised the jetties, frankly it admitted that Captain Eads had received only a fraction of the agreed price. Then, declaring that no reason was known why Congress, in 1874, should not have left the building of the jetties to its "own agents" instead of handing it to a civil engineer, except that Captain Eads had promised to do the work without payment unless the desired result was secured, they recommended that he be sent, not the million and three-quarters due him for work done, but two hundred and fifty thousand dollars as an advance to be used strictly on further operations.[23]

In a white-hot fury Eads railed back at the board, reviewing its report: "Of what avail would that be? Would it serve to relieve me from the mountain of debt which I bear? Would it pay, even to a small extent, the contractors whom I have employed, and who have devoted their whole fortunes and years of labor to this great work?"

Galled by the board's complacent credit-taking for the jetty venture, he lashed at their implication that he had been simply carrying out a system of improvements devised by the commission of engineers in 1874: "I urged this plan . . . I made a formal proposition to Congress . . . and offered to guarantee its complete success, long before the Commission of 1874 was ever thought of. . . . If I had not urged the jetty system with all the ability I could command, the usual agents of the government . . . would, in all probability, now be digging in the sickly marshes of Louisiana the canal recommended by a prior Commission."[24]

A furor in favor of and against James Eads in his insistence upon remuneration for the work he had so far done rose all over the country. Letters, pamphlets and newspapers lauded or condemned the jetty builder: he was a disinterested public benefactor imperiling his credit and reputation for the public weal, or he was a crafty opportunist using a flimsy pretense of aiding commerce in order to filch the public treasury. And as a rat-tat behind all this sounded the persistent sniping of Humphreys's brigade: Sea worms, teredo, were eating the willow walls, nibbling at them day and night, with the fanciful bar looming bigger at a new and impressive distance. Some of the press took up this patter.

In mock sympathy James Eads, who had often explained that the teredo would have considerable trouble getting at the mud-plastered, rock-flanked willows, remarked in a letter to one newspaper that "in addition to the distress worms had caused" among the Army engineers and a few journalists, they had had to relocate the bar again, this time five miles out to sea![25]

While darts flew in all directions, vessels from the whole Atlantic came and went through the jetty channel. Large Southwest Pass was wholly abandoned except by fishing smacks and a few small coasting schooners, and the dredges that had formerly worked there the year round had stopped months ago. "To realize how much the jetties have already done for New Orleans," presently wrote a special correspondent of the New York *Daily Tribune,* "one has only to sail along the river front of the city, where I counted last week no fewer than one hundred and twenty large square-rigged sailing vessels and eighteen ocean steamers. Fully four fifths of these ships come from foreign ports."[26]

What would Congress do for him when it convened? Would his debts ever be paid? Would his friends be ruined through him? Winter was setting in at South Pass, and the concrete caps not yet completed—could money be raised to go on with them? These queries tormented James Eads as he journeyed up the river in December, 1878, to St. Louis, and as he rode through the bleak streets. He could do nothing himself to aid the jetty finances, for he had lost his whole fortune on the Bridge that hung like a phantom in the gray air, the Bridge that no locomotive train had crossed for more than a year after it was opened for service. It was to be sold under the auctioneer's hammer from the steps of the courthouse on December 20, to satisfy a suit of the bondholders, a sacrifice to the railroads that had boycotted it into insolvency.[27]

His visit home was like a swiftly changing dream. Eunice and Adelaide were full of secret Christmas plans and shopping. Jacob Stiel, the gardener, had new triumphs in winter flower-growing to show off in the glass house. There was a glittering Christmas tree, the happy voices of Eliza and Josephine's children, a turkey dinner. There was a stream of callers, exchange of gifts. Cake and wine were passed, stories told . . . the jetties, what were the storms doing to the uncapped walls? There were greeting notes to answer, dinners and theater parties to attend . . . the Bridge had sold for two million dollars to Charles B. Tracy, acting for Anthony J. Thomas, of New York. Anyway it would serve the valley.[28] The jetties, what would Congress do about them?

While the holiday gaiety still rang all around him, James Eads, along with the other directors and the officials of the State Bank of Missouri, was called into court because the bank, like many others over the country, had suffered from the strikes and depression until it was compelled to close its doors. Shocked headlines glared from city newspapers and crossroads journals. Eads of the gunboats, of the big Bridge, of the jetties, to be hailed before a Grand Jury! A while later there were to be new headlines: CAPTAIN EADS EXONERATED.[29] The other gentlemen involved enjoyed a screen of comparative obscurity, but James Eads had long been favorite banner-line material, whether for his weal or his woe.

January, 1879, with old bills and new falling about him like snow—what would Congress do about the nearly two million dollars due him? He faced his creditors with the uncertainty of his position, he told them of his hopes, of the boon that the jetties had already proved to the valley, of their peril if left unfinished.

He managed to extend his credit, and was soon back at Port Eads watching the mammoth concrete caps grow seaward. They made the willow walls as strong as fortresses, but they were posing a new problem: the storm waves that once lost some of their force by sweeping over the jetties now raced along outside them, gouging the rubble, scooping it up here, dropping it yonder. Spur cribs would have to be built at intervals outside the walls to prevent this, and where was the money for them coming from?

Impatiently James Eads counted the days off against the damage of the racing waves. How many more days, how many weeks? In March he was informed that Congress had provided for about one-half the amount he had asked. This would leave him still nearly a million dollars in debt, but he could go on now building the spur cribs to prevent further damage from the waves.

He set these cribs a hundred feet apart, jutting them from the outside of each wall, and loaded them with broken stone and ship ballast, "dirt from the bluffs of Cape Town, gravel from the Thames, granite from Rio de Janeiro, sand from France and Spain."[30] Now that everything was going so well, he could accept an appointment of consulting engineer to the state of California that had been offered him. He promised Governor Perkins that he would examine and report upon the Sacramento and other California rivers, clogged with detritus from hydraulically operated gold mines, when he could find time.[31] But not for anything would he miss watching the Mississippi now as it carried out with freshet-heightened vigor its part of the jetty pact, shoveling sediment from its bed and tossing it off with the deftness of age-old practice.

By July 10, 1879, it had developed a channel with a middle depth of thirty feet through the jetties at average flood tide, a greater depth of water than had ever been required by any ship entering New York harbor, and seven feet more than the low-tide depth at Sandy Hook, one of the principal water entrances to New York.[32] The jetties were completed, the bar at the head of the pass entirely removed. The whole work had been accomplished in four years, and at a cost to the government of $5,250,000, little more than half of what Congress had once been about to spend upon *starting* the proposed Fort St. Philip canal.[33]

The country on the whole was jubilant, hailing the jetties as one of the most courageous engineering exploits ever attempted. "Genius, persistence and practical skill have seldom won so great a triumph over the forces of nature and the prejudices of men," a New York newspaper declared.[34] But a few diehards croaked on. "We lack faith in the story that comes from the jetties," the Cincinnati *Commercial* held out manfully against the general ovation.[35] And six months later the Memphis *Avalanche* of December 13, 1879, was to crash down with this: "It is a popular belief that the Eads jetties are a success. . . . The influence of the press has been mainly exerted in behalf of this stupendous fraud. . . . When the cash is all expended, and the contractors can see no prospect for any further subsidies, the dredge-boat will be broken up, the materials sold for old iron and firewood, and the famous jetty channel will be allowed to fill up with Mississippi mud."[36]

For all that, the largest ships in the world were sailing in and out of South Pass, eight hundred and forty steamers alone during the year of the jetties' completion. In the same period twenty-six times as much export went out through the Mississippi mouth as did in the year that the work was begun. The very same businessmen who had sent their representatives to Washington with desperate prayers that their port be saved from the havoc which the jetties would surely wreak upon it, now computed that the saving in freight on a single year's export of cotton would pay the entire jetty cost.[37] Moreover, the shipping of valley grain through New Orleans was trebling—wheat and corn vessels no longer had to sit stranded in Southwest Pass, their cargoes molding in the humid heat, until a freshet came to lift them over the bar. Insurance rates on these perishables had dropped, the price of bread was cheapened over half the earth. Immense land and floating grain elevators were being built at the crescent waterfront. New Orleans, raised from eleventh to second place as the country's export point, was taking its place as a great port, and there was talk of celebrating this fact with a World's Fair.[38]

The once vilified "outsider" was now adopted by the hospitable city, and the plaza at the foot of Canal Street named Eads Square for him, but he could not be pinned down long enough for many social honors to be showered upon him. The builder of the Mississippi jetties was called here and there over the country, with little time to sit on balconies telling stories or reciting poems. But busy as he was kept, he was intent upon a scheme he had long nurtured to improve the whole Mississippi, and upon a plan that he had lately adopted to open its commerce to the Pacific —"Shreve's river," the Louisiana folk used to call it; "Eads's river" it was now.[39]

CHAPTER XIV

THE TRAGIC ISTHMUS

IN HIS first days at South Pass, whenever a stentorian ship whistle rode in on the Gulf wind past the thundering pile drivers, James Eads had pictured a succession of ocean vessels sailing up between the willow walls, pushing the brown water into a cleft spray, and a ceaseless line of steamboats floating down a deep-channeled, shoal-free river to meet them. Even before he had snaked the workbarges down between the reed-flanked banks of South Pass and moored them at the slimy shore, he had evolved a plan to jetty the overwide and shallow stretches of the whole lower Mississippi, at least, creating a safe year-round navigation depth from the mouth of the Ohio to the Gulf.[1]

Since the days of the earliest keelboats rivermen had grumbled about the shallows that cursed the river channel. After an especially trying low-water season their protests might grow so noisy that Washington would lend an ear to them, occasionally going to the length of making a small appropriation for a bit of channel clearing or levee patching. The slight results of this meager, haphazard work were usually wiped out by the first flood. In late years the pleas of steamboatmen had become angry and accusing. The annual commerce of the Mississippi and its tributaries, they pointed out, amounted to more than the billion-dollar yearly foreign commerce of the whole United States, yet the aid given the streams was grudging and insignificant compared to the government gifts of two hundred million acres of public land to the railroads that were weaving their network across the continent.[2] James Eads had taken some of these chorused complaints before boards of trade. He had pleaded with his townsmen, after the close of the Civil War, to hold a conclave of valley river interests in St. Louis, and when the Grand River Convention materialized from this, he had postponed his gunboat mission to Europe that he might make an impassioned speech before the convention in behalf of the neglected waterways, arousing his hearers to demand aid from the government.[3]

After a heavy flood, in 1874, disastrous to the lowlands, Eads followed with interest the investigations made by a new commission of Army engineers, but read their January, 1875, report with dismay. It recommended a federal system of high levees to replace the hit-or-miss state levees, which was splendid, as far as it went. That might, for a while, protect the lowlands so recently devastated, but it did little for navigation. It would take deep water, not tall banks, to give boats a safe road, and the very highest levees, if not scientifically placed, would not provide it. Going to Washington with his river jetty plan, James Eads appeared before a Congressional committee and stated that the heaviest floods could be controlled within the river bed if that bed were deep enough—it was in the wide, shoaled stretches that the water raged over the countryside.[4] The bed could be deepened by contracting the banks of these stretches and forcing the river to scour out the silt, just as he had advocated for the deepening of the river mouth. Congress was skeptical about his plan to control floods, for in a year of extremely high floods "a million tons of water may flow down the Mississippi."[5] His plan was not accepted, but debates about it droned along futilely for several years.

In the meanwhile, James Eads had turned to the valley people with his plan for a deep Mississippi channel, lecturing, writing articles, publishing pamphlets whenever he could find time from his work at South Pass. There was too much inertia in the country's attitude toward the river roads, he accused. The people themselves must come to realize the importance of facile transportation. "School boys should be taught that superior facilities for cheap transportation secured to Phoenicia, Athens, Venice,

Genoa, the Florentine Republic and Holland the commerce of the world," he said. "Each retained it until a rival became a cheaper carrier; and it is a notable fact that art, refinement, literature, history and eloquence attained in each State their highest development during this commercial sway."[6]

Still stagnation reigned while the shoals grew and boats crept to their destruction against them. The frustration of the rivermen erupted at a River Improvement Convention held at St. Paul in 1877, at the time that James Eads and his unpaid men were struggling to get a 22-foot depth at South Pass. Every grievance, every sensational navigation incident was aired, and one of them was quoted threadbare: Sylvester Waterhouse, a well-known economist, had recently left St. Paul by boat on the same day that one of his friends had sailed from a port in Ireland, both bound for St. Louis. The traveler from across the ocean reached St. Louis first, the river trip taking Waterhouse thirteen days, owing to neglected channels.[7] The convention made a vigorous appeal to Congress, but failed to impress that body with the sad state of the valley waterways. A dab of emergency aid now and then was all, it seemed, that the rivers would ever get.

James Eads came forth again to flail this wasteful patchwork, demanding that the lower Mississippi, at least, be treated as a whole, its entire regimen respected. In May of 1878, a year before the South Pass work was completed, he brought a bill to Congress offering to apply the jetty system to the whole river, wherever needed.[8] This improvement would cheapen the cost of freightage not only by water but by land, he believed, for steamboat competition played a large part in controlling railroad rates. The cost of the jettying would be paid over and over by the benefit, for, he argued, "The keynote of our national prosperity is sounded in the simple words, 'cheap transportation.' "[9]

As could be expected, General Humphreys sallied forth against this bill—Jim Eads, the insatiable, not content with having snatched the delta channel from under the very noses of the Army engineers, was reaching to clutch the whole inland river program, and no one knew how much more he would try to grab. Humphreys wrote to a Congressional committee, listing the awful results of the then uncompleted South Pass jetties and predicting that they would continue to add disaster to calamity. He ended his letter succinctly: "It is hoped that sufficient has been said to show that there is no reason for transferring to other hands the charge of the survey of the river now going on under the Engineers' Department for the improvement of low-water navigation."[10]

Congress toyed with the matter, giving a preoccupied ear to the lobbyists. In the autumn, Humphreys recommended that the government ignore the dubious Eads plan and spend forty-six million dollars for an amazing system of levees that would have a maximum height of eleven feet above the highest flood ever known in the Mississippi. Eads's notion that the bed of the river was soft enough to be scoured by the current was absurd, he contended, for the bed was not alluvial at all, but was of prehistoric marblelike clay. He knew, for he had investigated by letting down a tallowed plummet in many places, in seven of which particles of this blue clay had adhered to it. This had all been set forth years ago, he recalled, in his own work, Humphreys and Abbott's *Physics and Hydraulics of the Mississippi,* a lengthy report to the War Department, and it was just as true today as it was then.[11]

If James Eads was a nettling challenge to Chief Humphreys, the doughty General was no less a chronic affliction to him. No one else could so vex him. He made at his antagonist now, tooth and claw, shredding Humphreys's letter and parts of his noted report, ridiculing the idea that the bed of the lower Mississippi was of hard, unyielding clay—he had walked most of this "ancient" bed and sunk in it to his knees. He reminded the author that at one of the seven places where the "prehistoric clay" had been found by the plummet, later borings disclosed a modern cedar log a hundred and fifty feet down through soft alluvium, and he regretted that the General had failed to explain how the log got there. He professed to be awed by the prehistoric samples of one thing and another that had clung to the exploring plummet—he feared that the General was not reverent enough of them.

"When we reflect that each of these precious specimens," he commented, "was deemed to be the key to an unwritten record running back into the dim past, where azoic and palaeozoic cycles inclose the sublime genesis of the Father of Waters, we cannot fail to note the terse expressions with which, in such simple terms, 'Gravel, Clay, Sand or Mud,' these antediluvium treasures are recorded."

He marveled that a profound work on the physics and hydraulics of a giant river should base some of its weightiest conclusions on anything so dainty as the pickups of a tallowed plummet: "Moses, when he stopped on Mount Pisgah, might as well have tried to analyze the subsoil of the promised land by gazing at it afar off as for these gentlemen to tell anything about the mythical substratum of clay under the shifting deposits of the river by means of their greased leads." Blue clay, Eads said, happened to be one of the common deposits of the stream, found everywhere in old wrecks, on sunken rafts, on piled driftwood.

"A few years ago," he gibed his adversary, "the Chief of the United States Army Engineers, being equally well convinced that the steamboat pipes were, like the bed of the river, unyielding in their nature, and that they were too high to pass under the Bridge which spans the Mississippi at St. Louis, accordingly recommended a canal with a drawbridge . . . be dug around the end of the Bridge in the ancient geologic clay of Illinois, at a cost of over three million dollars."[12]

A defender of General Humphreys, Rossiter Raymond, came back at Eads in the pages of an engineering journal: "The Mississippi itself is not muddier than his arguments. Unfortunately, his exhibition of conceit and ignorance on elementary questions does not render his challenge . . . a weighty one. His cool audacity may impress some minds as the sublime confidence of genius. . . . Those who have watched closely the career of Mr. Eads, while they recognize his enterprise and perseverance, and make allowance for his arrogance of temper, are not ready to accept his sanguine declarations in lieu of facts."[13]

With the arrows of controversy darting past in all directions, Congress found that no decision could be reached upon the method of improving the Mississippi, and put through a bill, on June 28, 1879, providing that a commission made up of three military, one coast guard and three civil engineers be created to study the various methods suggested and choose one of them. James Eads, finding himself one of the civil engineers appointed, meant to see to it that his jetty plan be selected, and that with no more loss of time than he could help.[14]

Steeped though he was in this controversy over the deep-channeling and flood control of the Mississippi, James Eads was mulling the most stupendous project that he had ever considered: a means of giving Mississippi commerce a short route to the Pacific, doing away with the costly 14,000-mile trip from its mouth, around Cape Horn, to California and the Orient. He had worked out his plan a year ago, in 1878, and intended to launch it after the completion of the work at South Pass.[15] But it was not always possible for him to play only one game at a time or to see all the pieces on several boards. While he was still occupied with the last stages of the jetty building and with trying to get his Mississippi deep-channel program accepted, Viscount Ferdinand de Lesseps, builder of the Suez Canal, called an international congress of engineers, geographers and naval authorities, in Paris, to examine the question of an interocean shipway across the isthmus that joined North and South America.[16]

If something came of this meeting, all right and good, Eads thought. It was time that the problem of the American isthmus was solved. Never since Christopher Columbus found the unwelcome strip of land across his dreamed-of western route to the Orient had men been able to reconcile themselves to the unbroken stretch of double continent reaching from the top to the bottom of the earth. They "could not believe that the Almighty, in making the world, had worked on a plan so apparently repugnant to the interests of humanity."[17] Occasional adventurers had searched the long American east coast for a break in it.

Balboa, in 1513, had hunted in vain for a narrow place at Darien where, the natives had told him, the tides of the two oceans sometimes met. After Cortez had conquered Mexico he had proposed a crossing over its narrowest part, Tehuantepec, "to shorten by two thirds the route from Cadiz to Cathay."[18] Pizarro had considered the isthmus a personal frustration after he had taken Peru and begun shipping its gold and silver to Spain. To do away with the hazardous voyage around Cape Horn he arranged that at a given time each year a fleet of galleons should sail from Spain to the east coast of Panama and a fleet from Peru for the west coast. Between the two landing places he opened a road through the jungle over which freight could be carried by beast and river barge from ship to ship. When Sir Henry Morgan, the freebooter, in 1670, descended upon Panama to loot it, the jungle road had degenerated to a mule path. Nearly two centuries later, in 1848, American gold seekers bound for California by sea cut a new way through the path and reached Sacramento Valley ahead of the covered wagons that plodded across the bone-strewn plains. Commodore Vanderbilt had shouldered in with a faster route at Nicaragua, a fleet in each ocean and a paved road between them.[19] Now, for years, the need of an isthmian shipway had been a poignant topic in Washington, and exploring parties seemed forever setting out for some section of the 1,200-mile long isthmus, or their remants returning to tell of the hardships and dangers in the "merciless and impenetrable forests." In 1869, the opening of the Suez Canal filled the United States Congress with concern lest global trade be concentrated in the Eastern Hemisphere. A shipway at the American isthmus was held imperative, and surveys were ordered.[20]

Right there the matter had dangled for ten years when de Lesseps, his talents idling, brought together the Paris Congress in May, 1879. The delegates came with maps of prospective routes and with arguments for crossing the long strip at various places, Darien, San Blas, Panama, Nicaragua and Tehuantepec, but they received scant attention. The prestige and personal charm of the aging de Lesseps dominated the meetings to the point that only his scheme, that for a *tide-level* canal at Panama, could get a full hearing. Although one of his countrymen, M. Le Blanc, who had lived on the Isthmus of Panama, declared bluntly that if anyone attempted to build a canal in that deadly climate there would not be trees enough in the thick tropical forest to provide wooden crosses for the graves of the laborers, the Viscount remained fixed that Panama was far the most suitable place for a transitway—meaning that it was the only one where a sea-level canal, such as the one at Suez, seemed feasible. It would cost, French engineers had estimated, about two hundred and forty million dollars and would take twelve years to build.[21]

When the Paris Congress closed with a chorus to the de Lesseps chant, James Eads strode forward in America, challenging the practicability of a tide-level canal at Panama. A canal with locks would be difficult enough to create there, he said, where the boisterous tides often ran three or four times higher than in the milder sea at Suez, where mountains, the rugged Cordilleras, would have to be cut through and the uproarious Chagres River controlled. He wrote letters that were printed on June 28, 1879, in New York papers, the *Times* and the *Daily Tribune,* presenting a plan of a ship-railway across the Mexican Isthmus of Tehuantepec—a railway to carry ships, cargoes and all, overland from sea to sea.

Startling the public with his ideas was nothing new to him, but this time he had outdone himself. Ships carried on a railroad car! And Captain Eads perfectly serious about it! Eads had broken across these ripples with his announcement that the Mississippi South Pass jetty channel had been completed. Then he started to the West Coast to fulfill the first of his duties as consulting engineer for the state of California.

On a day in late July, 1879, James Eads arrived in the California capital,[22] after what had been to him a thrilling journey across the western plains and mountains. No stranger could enter Sacramento casually, so colorful was the aura of its still youthful history. Only thirty-one years ago, on January 24, 1848, a flake of gold had been picked up a few hundred yards below a mill owned by John Sutter, an adventurous Swiss pioneer. Sutter, who had secured his 99-square-mile tract less than ten years before from the Mexican governor of California and, with his party of six white men and small troop of Indians and Kanakas, had put up some adobe houses and surrounded them with an adobe wall, tried to keep the gold discovery a secret. But it had leaked out, his ranch hands had deserted his crops and cattle to hunt for nuggets, and presently the world swarmed in on him, trampled his fields, killed or scattered his stock, took possession of his land, and he had fled the onslaught, a broken man.[23] James Eads could recall as yesterday the countrywide rush for gold—he had had a tinge of the gold fever himself and had hinted in a letter to Martha in March of 1849, when his debts were weighing upon him, that he might go to California, but later assured her that his "wrecking" was doing so well that he would not have to fare so far to repair his fortunes.[24] Now Sacramento, the once crude Sutter's Fort and later rowdy mining village, was a beautiful town of nearly thirty thousand people, but it was deeply troubled. Floods had devastated the lowlands, making a thousand square miles of them a sea, costing millions in wealth and hundreds in lives. The overflows were heavier each year, the levees as nothing before them. It was the hydraulic mines in the hills that caused the disasters, the detritus carried down every stream to clog the river channels until they could not contain the inevitable spring freshets.

After he had seen Governor Perkins, James Eads journeyed with State Engineer Hall and Colonel Mendell over much of the valley. Ruin and eyewitness tales of destruction and terror greeted him on every hand: Whole buildings had whirled downstream, trees tumbling after them; steamers that had customarily puffed along to the sounds of music and dancing fought their way on the gale-slashed, swollen streams with refugees; bawling cattle had stood in water on the crumbling levees, Chinese farm hands, "burdened with packs and splattered with mud," had scrambled

A STEAMER IN TRANSIT

along the levee tops, screaming at the crowded rescue boats to take them.[25] More eloquent than these stories were the gagged rivers and silent tortured land. Eads wrote of them: "Once these streams were pure and clear; there were fields producing harvests, extensive orchards, and substantial houses. Now these are all buried in the mass of earth which the uncontrollable floods have deposited, and prosperity has given way to desolation."[26]

He rode with the party up through the scenic hills to the source of the havoc, the mines. Here was another world. Groups of men toiled earnestly as high-pressure jets of water bored into the gold-bearing rubble, washing it into the sluices—it was none of their concern that Sacramento and Feather rivers were filled fifteen feet deep with waste, or that the channels of Yuba and Bear rivers had been wholly obliterated. It seemed remote and unimportant to them that floods had covered the valley like a sea. Their business was to mine gold.

This mine waste must be dammed up in the hills, James Eads said, and each damaged river compelled to clean its own bed. The faster their currents were made, the better excavators they would be—all questions relating to the movement of sediment hinged upon the simple law of nature: the work accomplished would not exceed the force expended. And the force would be adequate if State Engineer Hall's plan to narrow the wide sections of the streams with brush mattresses were adopted. He would send a full report later, Eads promised Governor Perkins, and return now and then to see how the plan worked out.[27]

Free now for a while to give his full attention to his scheme for an isthmus crossing, James Eads told of his plan wherever he could. While the general public considered it wildly visionary, most engineers agreed with Eads that a ship-railway across Tehuantepec would be entirely practicable and, in general, far from a novelty. Transporting loaded boats overland was an old practice. As early as the year 427 B.C. the Greeks carried their war vessels, boats a hundred and fifty feet long and having three banks of oars, across the Isthmus of Corinth on a kind of railway that remained in use for four hundred years. The Venetians, in 1438, had transported a fleet over a railway that mounted a hill more than two hundred feet high. Fifteen years later Sultan Mohammed II, of Turkey, built a ship-railway behind Galata, at Constantinople, and transported thirty ships of war from the Bosporus to the Golden Horn, five miles. And Emanuel Swedenborg, in 1718, during the siege of Frederikshall, contrived a ship-road with such success that he was raised to the peerage. Ship-railways had been suggested for Suez, and around the cataract of the Nile. Here in America Dr. William Channing had proposed a ship-railway for the Isthmus of Panama in 1850 and several times since, and a Mr. Day, in 1865, had published several articles stressing the practicability of such a road.[28] A number of engineers, among them General Q. A. Gilmore and Major Charles Suter of the Army Corps, the United States Naval Constructor, Captain U. S. Hartt, and the Dean of Civil Engineering at Cornell University, E. A. Fuertes, believed that Eads, the daring innovator, was less fantastic in proposing a ship-road across Tehuantepec than he had been in designing bridge arches five hundred feet long or in manipulating the Mississippi torrent at the head of South Pass. They later came forward and declared his ship-railway entirely practicable.[29]

He had been for years considering different plans to open the Mississippi commerce to the Pacific, Eads divulged, and had come to the conclusion that a ship-railway was to the best interests of the river because it was the only means of crossing wide Tehuantepec, only eight hundred miles from South Pass, nearer the jetties by twelve hundred miles than Panama. It would be an advantage to the whole country to have the shipway so near. He had spent many hours designing the road, working it out in cold figures after his own methods, building it in fancy, lifting imaginary ships to it from one ocean, carrying them overland and letting them down in the other ocean. A ship-railway would cost far less and be completed many years earlier than the de Lesseps tidewater canal at Panama could—if it *ever* could. It would make the Mississippi, within ten years, commercially an arm of the

Pacific Ocean. Later on he hoped to take up seriously the matter of an isthmus crossing.

For a while his design for improving the Mississippi absorbed him. October, of 1879, was at hand, the yellow fever had abated in the South, the commission of engineers on which he was to serve could start from Cairo on their investigations. James Eads was impatient to be about this, enough time had idled by, years of it, while men talked and the river raged. As the government-hired steamboat floated along over the water that mirrored the clear autumn sky and bright-foliaged trees, he began his campaign against the two principal plans that had been set up as rivals of his proposal to jetty the river wherever it was overwide and shoaled. One of the rival schemes, the so-called "outlet system," drew his especial ire. The main feature of this scheme consisted of opening an outlet from the river, ten miles below New Orleans, into Lake Borgne, an arm of the Gulf, to divert part of floodwaters. Eads pointed out that this designed outlet would begin to shoal as soon as the high water was over, just as the accidental outlets, such as the Jump, had done, and a struggle would have to be made to keep it open. The river, he insisted, could be helped to take care of its floods. Perhaps he was too jealous for the Mississippi, certainly he was almost fanatical in his determination that not one drop of its waters should be robbed from it.[30]

Nor was he much less rabid against the system of extremely high levees put forth by General Humphreys. Such towering walls, he said, would hold the torrent well above the countryside, a terrifying menace. As the bed shoaled more, the levees would need to be still taller. Why not make the river deep instead of high? Anyway, both of these systems dealt only with overflow, neglecting the problem of navigation. They treated the river as a hostile thing, a national misfortune, instead of as God's great gift to the valley. Its waters were precious, they should be conserved. It should be helped to retain its floods in a deep-cut bed. At South Pass it had been forced by the jetties to scour its bottom twenty-two feet deeper, surely it could be coerced in other too-wide sections by means of scientifically placed brush or, more substantial, walls, to dig its channel down a few feet. As time went on it would need less and less help from levees to confine its floods.[31]

It was the intimate way in which he talked of the mighty stream that won over the other engineers, his acquaintance with each bend and point, current drift and sand bar, and his plausible reason for each phenomenon. The rest of the commission discussed the river theoretically, but James Eads had swum in it, trudged up and down the weary hills of its bed—the Mississippi was as familiar to him as a farmer's road to town.

It was not surprising that the report of the commission, made three months later, commented with poetic awe upon the river's "great length and other elements of wonderful and impressive magnitude . . . ranging over a broad plane of its own creation," for James Eads had written it, and in high spirits—the commission had discarded all other plans and chosen his. By way of a test of it, two of the worst sections of the river were to be narrowed by jetties. One of these stretches, twenty miles long, reached from Plum Point (which was about a hundred miles below Cairo), where the low-water depth was only five feet; the other, several hundred miles downstream from there, was the Lake Providence stretch, thirty-five miles long, with a six-foot depth. All levees would be repaired, but not built higher.[32] This was the program, but it would have to wait until Congress adopted it.

In January, 1880, before he had finished writing the commission's report, James Eads read a colorful newspaper account of a visit that Ferdinand de Lesseps was making to Panama. The cordiality of Panama's welcome to the canal builder was so profuse that it consumed the whole yearly income of the provincial government and cost much painful preparation. The dumfounded natives of the capital had been compelled under heavy penalty to scrub and whitewash their houses and clean their streets—"such an air of cleanliness had not pervaded this city of pigs and smells within the memory of the oldest inhabitant." Through this unwonted neatness de Lesseps had ridden on a splendid white horse, his personal staff and others making up a triumphal procession which ended at the horse races being staged for the occasion. There were banquets at which the viscount told over and over why he had insisted upon a tide-level canal: large ships of five thousand tons had become common, he said, and some steamship lines were actually building vessels of six thousand, and even seven thousand, tons—what could ships of such size do in a lock canal?[33] His hosts could only look at each other, unable to answer.

One day James Eads came upon a news item that was of sharp interest to him: Viscount de Lesseps, with his wife and the three of their eleven children who had accompanied him to the isthmus, were sailing for New York on February 6. Here was a chance to persuade de Lesseps to abandon his canal project in favor of a ship-railway at Tehuantepec. He met the Viscount in New York and tried to interest him in the exchange, but de Lesseps would not listen to it. Eads then tried to prevail upon him to design his Panama Canal with locks, so as to ensure its completion within a reasonable time and at a cost that could be met. "A tide-level canal or nothing!" the old Frenchman had cried dramatically.[34]

De Lesseps had gone on to Washington to ask for Congressional approval of his canal, for military protection of it, and for security from competition by any other isthmus-way. James Eads, determined to launch a more practicable project, went to Washington, too. There he found that Rear Admiral Daniel Ammen, U.S.N., was on hand representing a company that had a scheme for a canal across Nicaragua, which lay about halfway between Panama and Tehuantepec. Ex-President Grant, now two years out of office, headed this company. All three of the crossing advocates busied themselves laying out their campaigns or writing bills. Presently they were invited to appear before a Congressional committee and describe their plans.[35]

The Nicaragua route appears to have had the first hearing. The isthmus there was wider than at Panama, but was cut well across by Lake Nicaragua and the San Juan River, leaving but a negligible strip of land to canalize. On the flat maps this route appeared simple, but the lake was more than a hundred feet above the Pacific, and the usually sluggish San Juan could be turbulent and dangerous in the rainy season.[36]

Ferdinand de Lesseps, seventy-four years old but looking no more than fifty, stocky, erect, alert, stood before the committee on March 8, 1880. Speaking through an interpreter, he said that for twenty years he had been convinced that a tidewater canal was the only solution of the American isthmus problem. The canal he proposed would run near the Panamanian railroad, would be 28 feet deep, 150 feet wide at the surface, less than half that width at the bottom, and would run more than two miles out to sea where it would meet a natural 28-foot depth. He planned to dam the Chagres River with a wall 120 feet high, and to tunnel through the Cordilleras. He had once thought that it would take eight or even twelve, years to build the canal, but he believed now that by using a large force of men he could complete it in six years; and his earlier estimated cost of $240,000,000 had been trimmed down to $168,000,000. He was quite complacent about the venture, speaking of it almost as though it were an accomplished fact.[37]

On the following day, March 9, James Eads stood before the committee. Sixty years old, slight, straight, studied of movement, he was as suave of manner and as inexplicably magnetic as de Lesseps. He was not impressed by the canal the Viscount had described, it was not as deep and little more than half as wide as the jetty channel at South Pass. Whether deliberately or not, Eads bore the attitude of an engineer commenting upon the hastily considered scheme of a promoter:

"I have no doubt that, instead of the narrow and tortuous stream that Count de Lesseps proposes to locate at the bottom of an artificial cut through the Cordilleras of Panama, engineers could give commerce a magnificent strait through whose broad and deep channel the tides of the Pacific would be felt on the shores of the Caribbean Sea, and through which the commerce of the next century might pass from ocean to ocean. . . . The only limit to the possibility of an engineer's profession lies almost wholly in the *cost* of the work which he proposes." He went on coolly, but dramatically: "Should there occur the need of a tower so high that it shall penetrate the regions of eternal snow, or an arch so great that its span may be measured by the mile, or a tunnel through the broadest base of the Rockies, or a railroad that shall transport, entire and uninjured, the grandest of the Egyptian pyramids; or a channel through Darien big enough to disturb the Gulf stream and alter the pulsation of the tides; you may be sure that each and all of these things, and much more, are within the possibility of his profession, if you will furnish the money to pay for them. The engineers of M. de Lesseps have agreed upon the sum of $168,000,000 as the ultimate cost of the work, exclusive of interest during the construction. . . . I believe that the estimates for the construction of the Suez Canal were forty millions, while the actual cost was upwards of ninety million dollars."

The Suez enterprise, Eads proceeded, was not comparable to this one. At Suez the tide was only six to nine feet high, while at the Pacific side of Panama the tide sometimes ran twenty-four feet high; at Suez the mean annual rainfall was less than two inches, but at Colón it was a hundred and twenty-eight inches. The turbulent Chagres River would cross the Panama canal five times in a distance of a little more than five kilometers, and the frightful rains that prevailed for half of each year would wash an incalculable amount of earth into the cut. If the canal were completed as small as proposed, it would soon be outgrown by shipping and could not be enlarged except at enormous cost and serious interruption of traffic.

Abruptly turning to his own project, he said that a ship-railway could be built for one-fourth the cost of a tide-level canal, and in one-third of the thirty years he believed that the de Lesseps canal would take to build. Over his ship-railway, which would run from Salina Cruz on the Pacific to Minatitlán on the Coatzacoalcos River, near the Atlantic not far southeast of Veracruz, steamships of the largest tonnage could be moved, with their cargoes, safely at four or five times the speed possible in a canal. The capacity of the road could easily be increased, without interruption of traffic, to keep up with the growing size and number of ships.[38]

To his wondering audience he then presented several drawings of the proposed ship-road. A basin three thousand feet long would be excavated at each shore, he explained; a ship-cradle, or car, would be backed down an inclined road to the sea end of the basin by a stationary engine to receive the ship. The vessel would be floated in from the harbor over the cradle from which arms-and-blocks would rise against the hull and secure it. The stationary engine would then pull the ship-laden car out of the water, two powerful locomotives would attach to it and draw it to the basin at the other ocean, where it would be let down into the sea end by another stationary engine and the ship floated off.

Ships were often lifted, without unloading them—it happened at nearly every drydock in the country. Nor was it an innovation in America to carry a loaded vessel overland. Forty years ago canalboats were regularly hauled across the Allegheny Mountains in Pennsylvania, from one canal to another, on a crude railway, and canalboats were now being transported by rail from the Potomac and lifted thirty feet to a canal.

His railway, he went on, would consist of twelve rails. Two locomotives would pull the burden along on four outer rails on each side, while tenders carrying coal and water would be hauled on the four inner rails. There would be twelve hundred wheels under the car, each with a strong steel spring over it. The ship would thus be carried on "a multiplicity of springs," and jolting almost entirely eliminated.

"I have no doubt," he stated calmly, "that at a speed of twelve miles an hour you would scarcely see a sign of motion on a glass of water standing on a table in the ship cabin."[39]

There was surprise among his hearers, but no incredulity. Few could listen to this resolute man and not be swayed by him. The meeting adjourned, as there was much for the legislators to consider.

James Eads hurried to New York to see his stepdaughter, blond, plump; gay Adelaide, who had recently married General J. G. Hazard, off on the steamer *Baltic* for England. He returned to Washington and on March 13 was again before the Congressional committee with his isthmus-crossing plan.[40]

Meanwhile many doubts and fears had arisen in the minds of the legislators about the ship-railway: the ship might crack in two on it—a vessel could not bear much change of position. Besides, a ship out of water naturally tended to fall apart. Meeting these qualms, Eads reminded his critics that he knew, from raising sunken boats from the river bottoms, about what the hull of a vessel could stand. He had built boats, was familiar with ship construction, and was satisfied that any vessel that was fit to put to sea could be carried safely on the railway. There was far more give and play in a ship hull than was generally believed—all the materials of which it was made were elastic.

"I doubt if there is anything in nature so hard that it will not change its form under pressure to some extent before it fractures," he said. "In a testing machine I have shortened a small column of limestone twelve inches long . . . nearly a quarter of an inch without fracture. Its length was restored as soon as the pressure was removed. . . . It is one of the rules of shipbuilding that a ship shall be so designed that if it is supported only at the ends it will not break down in the middle, and if supported only in the middle it will not break down at the ends."[41]

As for the fear expressed by some that the earth might crumble under the tracks when subjected to the weight of a ship, James Eads showed that the pressure on each square foot of earth under the largest loaded vessel would not be more than that of a six-foot column of stone a foot square, which, in turn, is much less than of the walls of a one-story brick or stone house, and utterly insignificant when compared with that of the Washington Monument. A man weighing a hundred and eighty pounds, using a crutch, exerts more than six thousand pounds pressure to a square foot with the crutch, yet it barely leaves a mark in the dust. Then there was the popular notion that only the pressure of the sea on the outside of a ship kept it from falling open—that, Eads said, was absurd. In a storm some part or other of the hull was, intermittently, well out of water. Ships were built to withstand tremendous strain when plunging and twisting in a high sea.

An earnest civil engineer, John M. Goodwin, was on hand to insist that the ship be carried in a caisson of water on the railway. In a cradle, he feared, it would collapse and lose its very semblance to a vessel, oozing into the corners and crevices of the car "as if made of putty." He enlarged this graphically: "You might compare a ship to a ripe cucumber which conforms exactly to the shape of the ground." As for the contention of Mr. Eads that an ocean vessel hull was often partly out of water, a ship did not, Goodwin insisted gravely, "have the impetus to throw itself out of the water like a whale."[42]

Blustery March had merged into April as the hearings for and against a Panama canal, a Nicaragua canal, and a Tehuantepec ship-railway with or without a caisson of water in which to carry the ships overland went on day after day. Witnesses favoring one or the other bobbed up from all over the country. All kinds of side issues edged in to consume time, even a discussion of where Panama hats actually were made. It was the middle of April when James Eads made a statement of the type of aid he wanted from Congress.

He proposed to form a company with a capital stock of seventy-five million dollars to finance the building of the ship-railway, but, owing to the magnitude of the work, he asked that the government, simply as a *loan,* guarantee the payment of six per cent annual dividends, for fifteen years, on two-thirds of this stock. None of the guarantees would be in force until the road had been tested by the carrying of a loaded ship over ten miles of track, and the lifting of the ship from a basin and the lowering of it back.[43]

The committee, thoroughly won over by James Eads's straightforward arguments, discarded all canal plans, and recommended the Tehuantepec ship-railway. A bill to provide a charter and financial aid for this railway passed in the House of Representatives, but failed in the Senate. This failure would lose him considerable time, Eads regretted, but it was not too discouraging. By another session of Congress the novelty of his plan would have worn off and there would be less suspicion of it.[44]

James Eads and Ferdinand de Lesseps had sparked the old and desultory discussion of an isthmus crossing into a seething controversy, a bitter four-sided struggle between supporters of a Tehuantepec ship-railway, advocates of a Panama canal, promoters of a canal across Nicaragua, and a large section of the voting public that believed the only sane thing to do was to dismiss the whole high-flown business and develop a self-sufficient, stay-at-home, mind-its-own-business nation.

Views casual or desperate were given forth by statesmen, engineers, bankers and boatmen. Arabesques of fantasy curved through the patter of dry facts and opinions: Marco Polo's route had opened the riches of the Orient to Europe, now a way through the isthmus would turn their flow to America; Kansas wheat would soon replace the rice in every Chinaman's bowl. There were allusions to the ancient continent of Atlantis, to the supposed visits of the fleets of Solomon and of Hiram of Tyre to the New World.[45] There were prophecies golden and somber. Echoes of this medley were to sound down the years.

CHAPTER XV

A SHIP-RAILWAY

WITH his heart set upon a ship-railway because it could be ready for use within a third of the time needed to build a tide-level canal at Panama, and upon a crossing at Tehuantepec because it was nearly next door to the mouth of the Mississippi, James Eads prepared for a trip to Mexico to seek that country's consent to his project. He was going to take Eunice with him, and Adelaide, who had returned from England. And there would be his lawyer, his secretary and two civil engineers, one of them Elmer Corthell, his resident engineer at the building of the jetties.[1]

The party sailed from New Orleans in the middle of November, 1880, on the steamer *Whitney*, floated past Port Eads to the Gulf, landing at Veracruz on the 21st. Eads, always refreshed by an ocean voyage, uneasy sailor though he was, enjoyed with his never-failing zest the trip over the eight-year-old railroad that wound up through the hills to the Capital which sat seventy-five hundred feet above the sea, ringed around by snowy peaks and rugged crags. In all his travels he had seen nothing more picturesque than this well-planned city, part historic, part new, enthroned on its ancient site, its distant street ends reaching for the mountains, an avenue of giant cypresses, dating from Aztec times, leading to the national palace that occupied the very ground on which the palace of Montezuma had stood. He settled his flock in the Hotel Iturbide, once the residence of an early Spanish Emperor of Mexico, Agustin de Iturbide,[2] and began casting his lines for no lesser catch than the President, General Porfirio Diaz.

The President was not surprised that a Mr. Eads from the United States wished to talk to him about building a railroad in Mexico, many American gentlemen, notably Mr. Jay Gould, had sought him out before for this reason. The railroad fever in his country had spread "beyond description." But this was a different kind of railway, a marine rather than a land enterprise, a road to carry ships from sea to sea, as intriguing a proposition as though M. de Lesseps offered to move his Panama canal project to Tehuantepec. General Diaz gave the American engineer immediate audience and received him cordially, apologizing for the confusion of his household. He had been moving his effects from the palace so that his successor in office, General Gonzalez, might move in. There was an inconvenient ruling that a Mexican president could not succeed himself—well, that might be changed in good time. About this ship-railway, he had heard of it for some months. If it could be achieved, Mexico would doubtless welcome it. Progress was his country's newly established policy.[3]

James Eads explained the prospective road and its significance. His swarthy part-Indian host fixed him with a shrewd, level gaze, finding the projector of this strange railroad unusual himself, so slight and fair, his eyes alight with enthusiasm, his manner that of "frank and exquisite courtesy"[4]—quite a contrast to the dark, Oriental-featured, soft-voiced Mr. Gould,[5] who ignored everything said to him that he did not care to hear. Diaz admitted that he was much impressed with the description of the ship-railway. He would recommend it to his Cabinet and leaders in Congress, whom he would arrange to have Mr. Eads meet. And they might ask a favor of the American guest—the harbors of Veracruz and Tampico needed improvement, perhaps such as had been given the mouth of the Mississippi.[6]

There followed a rapid series of conferences in which Eads set forth his plan to dominant Mexican statesmen. They were openly pleased with the straightforward proposal made to them. The ship-road, they saw, would virtually give their country a new shore line and a key position in the world's commerce. They were willing to grant the half-mile wide right-of-way asked of them, and would add to it a million acres of tax-exempt public land. But when the question arose of sharing authority over the road with the United States, whose territory it did not touch, they ruled it out emphatically—the door that the American Congress had lately shut against the ship-railway must remain closed.[7]

Taken somewhat by surprise, James Eads came back with equal firmness, making it clear that he had no idea of creating an isthmus-way that was not partly under the control of his own country, or of throwing the Mississippi mouth wide to various powers. Indeed, he intended bringing his request for an American charter before the United States Congress shortly. None of this, he said, would affect the certain benefit that the shipway would prove to Mexico. But the Mexicans were uneasy. They had not forgotten, even though they did not mention it, their war of 1846-1848 into which, they believed, they had been shouldered because their strong neighbor to the north coveted the coast of California and anything else handy.

Steering a hairline course of diplomacy, James Eads used his utmost powers of persuasion. In two weeks of skillful chess tactics he managed to allay suspicion and to secure an outlined charter in which there was no protest against aid from any country. It agreed, moreover, that the United States could regulate tolls and send mail, warships and other government property across the isthmus toll-free; it gave exemption from import duty on all equipment necessary to the construction of the railway, and from export duty on all money sent out of Mexico in relation to the work; and it pledged to co-operate with the United States in protecting the route from aggression. This grant would, however, have to be confirmed by the incoming government.

Now James Eads brought up one more request: he needed aid that only Mexico could give in surveying the isthmus. This was unstintingly accorded. A commission of fifty men—engineers, soldiers, laborers, with the noted engineer, Francisco de Garay, at their head—was assembled. The President detailed a man-o'-

war, the *Independencia,* to carry Eads and his now considerable party to their explorations.[8]

They skirted down the east side of the isthmus, a sleepy, primitive world, its indolent people, of mixed races, living in thatched huts and all possible idleness. After examining the coast, they turned into the Coatzacoalcos River and went up that stream some eighteen miles to Minatitlán, where the ship-railway would commence. Primitive river craft pulled out of the way of the big vessel, the startled boatmen peering sullenly back at it. From a masthead of the ship James Eads looked over the surrounding country. There were no hills in sight, but there were lush, impenetrable forests of mahogany, pine and cedar through which, over an ancient narrow trail, pack animals still bore freight from ocean to ocean, male travelers following on horseback, women carried on the backs of peons. Hills that rose farther inland smoothed out at last to the cactus-grown plains of the Pacific side, the home terrain of sturdy aboriginals, intelligent, well housed and cleanly.[9]

His trip to the isthmus, besides the thrill of its novelty, added to James Eads's conviction that it was the best site for a crossing. Of the one hundred and forty-three miles of his route, a dozen would be by the bay and thirty more by two rivers, leaving only about a hundred miles to be traversed by railroad, and all of it through dry valleys or over plains except for one short section of marsh.[10] The cordiality of the Mexican government had been encouraging, their generosity in sharp contrast to the turndown he had suffered from his own country.

After three weeks on the isthmus Eads returned to New Orleans full of his project. He now believed that his ship-railway could be built at one-fourth the cost of a Panama tide-level canal, and perhaps in one-fourth the time, that its maintenance and operation would be much less than that of a canal, that it could offer far lower freight rates, and that a ship would be less liable to accidents on the railway than in a canal. For the money needed to put through de Lesseps' venture, ship-railways could be built at four or five places across the long strip of land reaching from Mexico to Colombia.[11] In St. Louis he talked and breathed ship-railway, making an address about it at the Merchants' Exchange.[12] He went East to discuss it with other engineers, he sketched his plan for shipowners, wrote of it to journals and newspapers.

The reason that the world had been daunted for three hundred years by the slight barrier that closed ocean from ocean was, Eads declared, the tendency of men to cling to familiar ways. Throughout history they had blindly opposed innovations—almost within present-day memory they had resisted the introduction of steamboats, of locomotives, and of the telegraph. A supporter of the de Lesseps Panama canal admitted that in 1835, when he was fostering a railroad through Indiana, his opponents insisted that a man who believed that a train of cars could pass across the whole state "must be a monomaniac."[13] A few years later the chairman of the English Stockton-Darlington Railway had told Parliament, after the road had daringly begun to use a steam locomotive instead of horses—but, of course, with a signalman on horseback galloping ahead to warn the populace of their danger—that no further improvements in railroading were likely possible, except, perhaps, "a high earthwork bank on each side to prevent engines from toppling over."[14]

So intent was James Eads upon his project that his zeal pushed back the fatigue of what had been, even for him, a strenuous year. His work had grown so heavy and farflung that he persuaded Estill McHenry, Josephine's husband, to take the post of secretary and lift the weight of all but the ship-railway routine from his shoulders. McHenry moved his family to the Compton Hill mansion to be with Eunice, who was lonely after the marriage of Adelaide.[15] Listing some of the things that had occupied him in the twelve months of 1880, James Eads wrote:

"I inspected the River Danube for about eight hundred miles of its course, and investigated the cause and extent of the frightful inundation at Szegedin, Hungary, which involved an examination of a hundred and fifty miles of the Theiss River. I also examined the Suez canal to familiarize myself more thoroughly with the question of a canal across the American isthmus, having previously visited the Amsterdam ship-canal and the one at the mouth of the River Rhone. As a member of the Mississippi Improvement Commission I also aided in perfecting plans for the improvement of that river, and the preparation of its report now under consideration before Congress. Within this time I have thrice visited the jetties at the mouth of the Mississippi, besides my visit to the City of Mexico, Tehuantepec and Yucatan. . . . I have also at the request of the Mayor of Vicksburg twice visited that city during the last year to examine its harbor with a view to its improvement."[16]

And a fine brawl he had had with the commission over the Vicksburg trouble! Four years before, in 1876, the Mississippi had broken through the peninsula which reached up from the Louisiana shore and stretched a long finger across the front of Vicksburg. In no time the wharves, warehouses and grain elevators were left standing on the edge of a lake in which the remains of the old peninsula appeared as an island. The harbor began to silt up, deteriorating rapidly. The port was alarmed. It was about to be shut off from the river, its picturesque bluffs rising above a mud flat. The majority of the River Commission wanted to dredge a channel through to the harbor. James Eads said that the river would have none of this local patchwork, it would dump silt into the cut as fast as a dredge could remove it. Instead, they must consider the changes, contributing causes, that had been taking place a hundred miles above and below the port. The stream would have to be manipulated until it undid the damage it had caused and once more washed deep and brisk against the face of the wharves.[17]

It was in this same year, too, that James Eads, on a second trip to California, had, on August 11, spoken before the San Francisco Chamber of Commerce on the subject of the ship-railway, facing a very prejudiced audience. Californians believed that an isthmus crossing of any kind at any point would dissipate a large part of their Oriental trade, parceling it out among East Coast and southern ports. Reeling off figures about their imports and exports, Eads carried his hearers to a conviction that the benefits of a crossing, especially if at Tehuantepec, would immeasurably overshadow any adverse effects.[18] Later they petitioned Congress to support the ship-railway. On his way back from California he had "visited the wonders of Yellowstone Park, crossing the Rocky Mountains in that excursion six different times."[19] And traveling in Yellowstone was not easy, much of it having to be done on horseback or in coaches, with pack animals trotting alongside carrying equipment for the night's camp.

With no more magic than his reaching interest and boundless energy, he had traversed continents, crossed oceans, climbed mountains, sailed rivers, worked, coaxed, fought, played. For James Eads the year 1880 had been just another twelve months filled to overflow with living.

In January of 1881, he came back to Washington with his ship-railway plan. The promoters of the other routes and schemes had gathered there, too. The competition was grim as their causes again lined up before a Congressional committee. For once, the Army engineers were not arrayed against James Eads, but several Navy officers belabored his plan soundly—to a Navy man the place for a ship was in the water, not perched on a railroad car. One of them, Captain Seth L. Phelps, a veteran of the Mississippi ironclad fleet, represented the Nicaragua canal company and was fortified by a concession from the state of Nicaragua and an opinion from Mr. W. J. McAlpine that it would take a ship six days to be carried across Tehuantepec on Eads's railway. The American representative of the Panama canal company, ex-Secretary of the Navy Richard Thompson, a genial, gray-haired, black-eyed man, argued gracefully that the United States would do well to foster the de Lesseps canal so as to share control of it, but he alarmed the legislators by dwelling upon the dam, a hundred and twenty-four feet high and a mile long, that the French engineers thought would be necessary to keep out the mad waters of the Chagres River—the wall would hold aloft six hundred million cubic feet of water, which it would allow to flow gradually into the canal.

James Eads saw the committee turn in some relief to his ship-road. He had never been more confident than he was now, standing there with his charter from Mexico and letters from the most eminent ship constructors of America and England warmly recommending his plan. He disposed offhand of the opinion of McAlpine that it would take six days to haul a vessel across the isthmus on the railway, recalling that the same gentleman had headed a convention of engineers at St. Louis in 1867 that unanimously condemned the deep foundations and long arches designed for the Bridge as "wholly impracticable"—Mr. McAlpine could be wrong again. Eads reminded his hearers that since his first bill had been introduced nearly a year ago not one engineer of note had come forward to challenge the practicability of the ship-railway. He pleaded that Congress give him the approval he sought, and the financial guarantee loan. He stressed the flexibility of his railway to keep pace with the growth of ships, warning that "the vessels of today may be insignificant in size compared to those of twenty-five years hence."[20]

He had put his case with forceful simplicity. The committee was wholly won over to it. Reporting emphatically against the de Lesseps Panama canal and mildly against the Nicaragua project, they heartily endorsed the Tehuantepec ship-railway, saying:

"In considering this proposition, the question presents itself, *is a ship-railway practicable?* The evidence before the committee upon this point is overwhelmingly in the affirmative. . . . Some of the most able and prominent engineers in the world have declared in letters, which have been submitted to the committee, that the project of a ship-railway is in every sense practicable. . . . It is a noticeable fact that *no* engineer has appeared before the committee denying its practicability. . . . He [Eads] does not ask the government to advance the money necessary to construct the railway, as was done in the case of the great Pacific trunk lines, but simply that it shall guarantee the payment for the period of fifteen years of six per centum upon the par value of fifty million dollars of the capital stock of the company. . . . Wherefore your committee report the accompanying bill with the recommendation that it do pass."[21]

For all this, the bill introduced in the Senate, by Senator Vest, of Missouri, met such a storm of opposition that it was taken from the calendar and left to languish in the Committee on Foreign Relations. Frustration gripped Eads, his arms were pinned to his sides. It would take eight years to construct the road, and time was drifting. But in the end the ship-railway would win its fight against old and traditional methods, just as the jetties had. Another session of Congress would surely see his bill go through.

Meanwhile the Panama canal venture of Ferdinand de Lesseps had fared better than many had expected. James Eads, searching the newspapers for items about it, read that the Viscount, upon his return to France, had reported finding that his tide-level canal could be constructed with ease, and that its financial success would be enormous. "As for the salubrity of the climate, how unjustly it had been condemned by those who knew nothing of it!"[22] Continuing a series of dinners begun after the Paris Congress, de Lesseps had brought his scheme to the close attention of many influential men. Eads had doubtless had a good laugh over an editorial in the New York *Times* inspired by the earlier banquets: "The whole plan of the canal, from the oysters to the cigars," it bubbled, "reflected the utmost credit upon the *chef,* and by the time the third successive dinner was eaten it was felt that the canal was as good as cut. There were, of course, certain dyspeptic persons who maintained that there were difficulties in the way of cutting a canal through a range of mountains. . . . Mr. de Lesseps pointed out that thirty miles in a vertical direction were no longer than thirty miles in a horizontal direction, and that he would take a little more Chateau Lafitte. . . . Warming with his subject and the curry, the veteran canal digger proceeded to illustrate the route of the Panama canal with the aid of forks and knives—showing how it would start out with the salt cellar . . . and finally reach the peaceful expanse of the butter dish, that is to say, the Pacific Ocean. The company was completely satisfied and decided to begin it without fail at 8:45 the next morning."[23]

And begin it they did, although somewhat later. Having attracted a hundred thousand shareholders to the project, de Lesseps shipped equipment to the canal site and the work had been commenced in February, 1881, just as the promoters of the Nicaragua and Tehuantepec plans were making their second campaign before a Congressional committee.[24]

It must have seemed to James Eads that his gracious home on Compton Hill had become mainly the place where his baggage was packed for another trip, so constantly did his work and his curiosity keep him moving about. Early in April of 1881 he was on his way back to Mexico, having in his party Jesse Grant, engineer son of ex-President Grant. Although his father headed the Nicaragua canal company, Jesse stood firmly for the ship-railway. At the Iturbide Hotel, Eads found, as he expected, General Grant, who had arrived several days before on a mission in behalf of the Mexican Southern Railroad, which he was promoting. This would have been the time to bring Eunice and Adelaide, for Mrs. Grant had accompanied her husband, as had also Ulysses Grant, Jr. The two old friends, now of rival isthmian ambitions, planned to see as much of each other as busy men could, each now engaged in getting the ear of President Gonzalez. Eads was given a long conference, on April 13, with the Executive, who assured him that the promised charter would be passed upon in his Congress at the earliest possible moment.[25]

As James Eads went about the city, sightseeing, calling upon friends he had made during his earlier stay here, or making new ones, it seemed to him that he was forever cutting across a gay procession or getting caught up in it—this was Holy Week, a time not squandered here in fasting and sorrow as in other lands.[26] He visited the newspaper offices and explained his ship-road to the editors. Later press comment described him as energetic, simple but elegant, past sixty years old but looking years younger. "His sympathetic face has the pleasant and fine features of a man of the world," *El Nacionale* said of him. "In his gaze glows genius; in his countenance, goodness of heart; in his modest words a torrent of knowledge."[27]

He drove out the long beautiful avenues with the Grants, the mountains shifting a panorama ahead of them. He went over charts with the engineers he had set to surveying the harbors of Tampico and Veracruz, he read reports from his engineers at Tehuantepec. He had expected to spend a month at the isthmus, cross it to the Pacific side and sail up the west coast of the United States to fulfill promises to inspect Humboldt Bay and the mouth of the Columbia River, advising methods of improving their harbors. But if he went to the isthmus at all, he postponed the trip up the western coast to California and Oregon, for on May 28, immediately after the Mexican Congress confirmed his charter, he boarded a steamer at Veracruz for New Orleans.[28]

General Grant, who had succeeded in his Mexican errand, too, that of getting his railroad charter modified, was aboard with his family, the two parties making a jolly group on deck and in the dining saloon. And there was some earnest talk between the heads of the rival isthmus projects. Eads described his ship-railway in detail and showed the General the new bill he had drawn up to present to the United States Congress. Perhaps the influence of his son, Jesse, had much to do with it, but certainly the president of the Nicaragua company was losing confidence in his own venture and swinging to the side of the other. He told Eads that he could not support the ship-railway, but he would not oppose it.[29] On these pleasant talks under the blue Gulf sky was yet to hang a hateful controversy, between and around the two friends, kindled by Admiral Ammen, who could not, however, keep it long ablaze.

When the vessel entered the jetties, James Eads asked the captain to drop him off at Port Eads, and was surprised when he was told that he could not land there, nor at New Orleans until he had been through quarantine. Veracruz had been reeking with yellow fever when the boat left there, and all "unacclimated" persons would be held for inspection. Eads was frankly irritated at being listed as "unacclimated," as many times as he had walked in the very shadow of the yellow killer! Moreover, he was in a hurry.[30] There were Mississippi improvement reports to glance over, harbors on the West Coast awaiting him, and he had a special commitment in Canada—he had promised the Minister of Public Works to come to Toronto at the first possible moment to inspect the distressed harbor there.[31]

After a short pause in St. Louis, his bags freshly packed, James Eads was off to Canada. He found Toronto a beautiful city and an interesting port. Sitting at the head of Lake Ontario, opposite the mouth of Niagara River and between the rivers Don and Humber, it was an essential center of trade and travel. He boated around the bay, which was virtually a small lake off the large one, formed by a half-circling strip of land with a narrow entrance at the west end. It was this gateway between the bay and Lake Ontario that was in trouble, the growth of the land strip threatening to close it. The opening could be widened and deepened, James Eads said, by the very currents that were relentlessly filling it up, if dikes and breakwaters were built so as to direct the forces in their work. But the problem here presented some novel features of current and sand movements, and it would be several months before he could submit a detailed plan.[32]

There still lay the West Coast work to be done. Oregon officials awaited him anxiously, for the bar across the mouth of the Columbia River, obstructing navigation, had seemed more intolerable than ever after the glowing accounts of the new channel at the entrance to the Mississippi. Their river was important, too, the largest ships could sail the hundred miles upstream to Portland and far beyond, if they could get over the bar—the Columbia's river system had more than two thousand miles of navigable water. When James Eads came into the Oregon country it was as though he had entered a land designed for Titans. The Douglas firs rose two hundred and fifty feet high, the mountains were lofty, the upper Columbia tumbled in rapids and surged in millrace currents, tides pushed upstream a hundred and fifty miles, the mouth of the river was nine miles wide, the ocean raged stormily over the bar. It was superlatively beautiful, it was rich in promise. In time, Eads thought, the bar might be wholly conquered. He recommended that jetties, such as he had built at South Pass, be run out two and a half miles. He was not free to undertake this work, he said, he could only advise.[33]

He was tired when his western trip was over, and resolved to take a rest by going to England and doing what he could to persuade ship lines there to send their vessels across Tehuantepec when the ship-railway was completed, instead of through a Panama canal. It was the middle of July when he sailed from New York on a handsome new German Lloyd steamer, the *Elbe*, in company with Captain John A. Dillon, son of his cousin Eliza Eads Dillon, a St. Louis newspaper editor.[34] England had always relaxed James Eads, the country was so green and peaceful, its people so hospitable to him. And there was often Genevieve, his oldest stepdaughter to visit at Southampton where she looked after the home of her husband, John Ubsdell, part of each year, as he was now supervisor of the South Pass jetties and tied down at Port Eads.[35]

James Eads talked to shippers and shipbuilders at different ports, he looked at rivers and canals, he poked about London. He was surprised, but no whit displeased, that engineers and scientists made much of him. Sir John Lubbock, president of the British Association for the Advancement of Science, invited him to attend a session of that organization at York. There he found himself elected a member and called upon for an address about the Mississippi jetties and the proposed ship-railway. He protested that he was wholly unprepared, but his excuses were brushed aside. He consented to "occupy a half hour or so on each subject," and notices were quickly given out that Captain Eads, one of the greatest engineers of all time, was going to speak in the hall of the Corn Exchange directly after the president's address on Thursday.[36]

The size and eagerness of the audience that awaited him

A SHIP-RAILWAY

amazed James Eads. He could not understand why the jetties were considered so sensational an achievement—as for the ship-railway, it was merely the most obvious and simple means of getting ships from one ocean to the other, in spite of a narrow land barrier. On the other hand, a canal across mountain-spined Panama, with its deluges of rain and its roaring river floods, appeared to him a sensational project.

Looking out over his decorous British audience, James Eads, without script or notes, talked of the little-understood movements of current and silt, making sketches on a blackboard as he went along. With anything that produced friction, he said, even a fish net, set in or drawn across any part of a stream, the current would be relaxed enough to cause more than ordinary deposit of sediment beyond it. But if any means enlivened the current, the sediment would be borne along, and even more of it picked up from the soft bed until the load matched the velocity. He recounted his manipulation of the Mississippi torrent at the pass heads as though it were no more than pouring water from one jar to another. He was utterly casual over the fact that the largest Atlantic ships were sailing through a thirty-foot deep channel where, only six years ago, small fishing smacks had grounded in the mud. He made it simple and dramatic, and his audience forgot that, for the most part, they had expected to wait politely until he got around to the matter of the ship-railway.

The American isthmus, Eads emphasized, was about two and a half times as long as Great Britain, and commerce moving through a Panama canal would be compelled, in most voyages, to travel down and up its coasts two thousand or more miles out of the way. A crossing at Tehuantepec would save much of this wasted distance. He described his ship-railway design, mingling in a vivid pattern the engineering principles and his vision of its effect upon world trade. The subject gripped him and his hearers, bearing them along together. They could see the vessels pulled up from the sea and carried through the tropical jungle, with every wild creature frozen to silence as ship masts or steamer stacks moved by, they could feel themselves glide with the vessel down into the second basin and rock out to sea. The ship-railway was inevitable! The only wonder was that it had not been built years ago. James Eads, talking to scientists, not thrill seekers, was gratified when they spontaneously acclaimed his offhand address as of such scientific importance that it should be embodied in the association's records to preserve it.[37]

When he got back to the United States he was so rested that he took up a number of waiting tasks. Then, in late October, 1881, he sailed from New Orleans on the *City of Merida* for Mexico, picking up his son-in-law, John Ubsdell, at the jetties and taking him along. He found the Mexican capital in a great state of preparation for the feast of All Souls and All Saints, which was nearly at hand. Booths were being put up along the gracious avenue, the Alameda, where the customary gruesome articles, tiny coffins, skulls and tombstones (sugar ones for the children) would be sold. In another week the city was babbling with gossip, ex-President Diaz, a widower, had quietly married pretty Señorita Carmen Rubio.[38]

Several matters detained Eads in the capital and thereabouts, chief among them the surveys he was having made at the harbors of Veracruz and Tampico. It was November 24, 1881, when he went on down to the isthmus, where his engineers were still seeking the most direct route with the fewest curves and least elevation, and laborers were cutting a twelve-foot wide path through the thick forest. It was the middle of December when he got back to South Pass and took a tug for New Orleans. He had come home before he wanted to, he told news reporters, but the work that he was advising in Toronto demanded his attention.[39]

While James Eads had been at Tehuantepec, a report had turned up in the United States from Panama that splendid progress was being made on the canal. The confidence of M. de Lesseps had soared to such a height that he had lopped sixty-eight million dollars from his last estimated cost. Complete with docks, he published, the canal would cost only a hundred million dollars.

But in October, on commencing the long open cut through Mt. Culebra (he had abandoned the idea of a tunnel), the deepest part of which would be five hundred and thirty-eight feet and the average depth two-thirds of that, the first of a series of colossal difficulties was at hand. The Viscount, it developed, had accepted only the most flattering geological surveys, and one of these showed Culebra to be covered with a light soil through which the mountain stone protruded. Actually, the stones thrusting out from the soil were dolomite boulders lying embedded in a thick deposit of clay. The heavy rains splashing in cataracts down the sides of the cut would wash tremendous quantities of earth and stone into the canal.[40] Enemies of the Panama venture gloated over this, but the optimism of de Lesseps was undisturbed.

Advocates of the various routes and plans were filling the air with oratory, sounding the same cry that had heralded the transcontinental railroads: "On to the Orient!" Each group had to face two sets of opponents, setting off controversy within controversy. One of the wordiest was that which centered about General Grant. After his trip from Mexico with James Eads, Grant had advised Admiral Ammen, of the Nicaragua canal company, against pushing his project, as de Lesseps had a large popular backing in France, and English financial support was waiting for Eads to accept it for his ship-railway. Later the General had gone further, asking that his name be removed from the rolls of the Nicaragua company. Ammen was deeply hurt and blamed Eads for Grant's defection. Accusations and denials were leaping nimbly from the pages of newspapers and journals when the promoters of the several schemes came back to Washington in December, of 1881.[41]

Everything looked fair for the ship-railway. Much of the country had risen up in its behalf. In the past year thirty-four petitions had been sent from various states, thirteen from Mississippi, six from Texas, four from Alabama, six from Missouri. Louisiana, California, Georgia and Tennessee had chimed in. And a Mr. Slayback had sent one all by himself.[42]

The Capital turned from its sotto voce surmises about the love affairs of Lily Langtry, its astonishment over Oscar Wilde, who, elegant in satin evening coat and knee breeches, lace ruffles foaming from his cuffs and curls dangling on his shoulders, had paused in a country-wide series of lectures on aesthetics, and gave its curious attention to a discussion of an isthmus shipway that was growing peppery on Capitol Hill.[43]

The committee debates, which were to carry on well into the spring of 1882, opened on a testy personal note. And directly Captain Seth Phelps, of the Nicaragua company, confided to a Select Committee from both Houses that Captain Eads, having had to go to England to find any approval of his ship-railway, had there "struck upon *Mr.* Reed, ex-naval constructor out of employment. Having since come to this country, Reed had been going about with Captain Eads to talk up the ship-railway, being presented as *Sir* Edward Reed, M.P., and I know not what other titles." Besides all this, Phelps complained, it would take a ship eight or nine days to ride across the isthmus on the ship-railway—a slowup of two days since he had reported on this phase a year ago.[44]

James Eads brushed personalities aside almost brusquely when he came before the committee.[45] He was full of certain drastic changes he had made in his ship-railway design: He had discarded the inclined track that ran down into the shore basins, and, instead, would have a caisson lift hoist the ships straight up the forty-six feet necessary. For this each basin would be supplied with a great steel pontoon, or caisson, which, to receive the ship, would be filled with water until it sank to the bottom of the basin, the cradle resting on its top, or "deck." When the ship had floated to position over the cradle, the water would be forced out of the caisson by powerful centrifugal pumps until the caisson rose so that the keel groove on its top fitted under the keel of the vessel. After the many rubber-cushioned supporting arms had swung against the hull, further water would be pumped out of the caisson, allowing it to rise, kept level all the while by hydraulic governors, to where the tracks on the caisson top met the tracks on shore. Now the locomotives would attach to the car and pull it and its burden away. The number of wheels had been reduced by half. The turntables, to be placed at five points on the route to eliminate too-sharp curves, would rest on pontoons floating in basins of water.[46] All eminent engineers who had studied the plan, Eads said, had approved it. Two noted English firms of contractors, one of them builders of the powerful hydraulic docks at Malta and Bombay, had volunteered to construct hydraulic lifts for his railway, guaranteeing to raise a loaded ship of ten thousand tons and place it on the land tracks in thirty minutes, in perfect safety, but he had decided to employ caisson lifts, as they were less costly. The English constructors had also guaranteed to transport the vessels overland, with no damage to them.[47]

This called forth again the old cries from Admiral Ammen and Captain Phelps that vessels cocked up on a car would be blown over, one after another, by the wind. While ships were thus being flopped off the railway into the jungle at an appalling rate, Sir Edward Reed, retired constructor of the Royal Navy, "happened to be passing through Washington," and, to the snorting disgust of Captain Phelps, "kindly consented" to appear before the committee. "It is perfectly impossible," Sir Edward testified, "for these ships to come to any grief from wind, because the resistance to hold the ship upright in her cradle on the tracks is, I think, very many times greater than the forces which keep her upright at sea. . . . I have searched in vain and cannot find any element of danger."[48]

As the debate droned on, popular interest flitted to other things. Washington was gay. The period of mourning for President Garfield, who had died in September from an assassin's bullet after two months of pain-ridden lingering, was over. Friends of President Arthur, office seekers, lobbyists and many who idly gravitated to the Capital, jammed the hotels. Against the crowds an occasional colorful figure stood out: Dr. Mary Walker, dressed in pants, Prince Albert coat and stiff shirt-front, "a light cane in her withered hand and her plug hat pushed back on her head"; and Robert Ingersoll, "the pagan," fascinating or repelling those who heard him lecture.⁴⁹ Gentlemen in long-tailed coats, their shiny high hats and gold-headed canes held in a gloved hand, ushered ladies, modish in tight-waisted evening gowns, from carriages to lighted doorways that opened to gusts of chatter. Sparkling chandeliers beamed down on tables, touching damask and silver with a pattern of high lights.

Not infrequently James Eads was the host of such an occasion. His growing popularity in the Capital was a matter of comment. "Socially Captain Eads was one of the most charming men who ever came to Washington," a New York newspaper afterwards observed.⁵⁰ Months ago Admiral Ammen, in a letter to the New York *Times,* had complained that Captain Eads was "engaged in using all those peculiar methods so well known to those having long experience in working up legislation such as he is now striving to secure. . . . Senators and Representatives are entertained with dinners, and costly bouquets and baskets of flowers are sent to their wives with the compliments of Captain Eads."⁵¹

It was nothing new for James Eads to dine his friends or to present flowers to ladies, but likely he was now making out his guest lists and dispatching his bouquets with the ship-railway somewhat in mind. One of his Washington dinner parties, conspicuously noticed in a New York paper, seemed to bear out Ammen's complaint:

"Captain James B. Eads gave a dinner, with twenty-eight covers, at Wormley's this evening complimentary to Señor de Zamacona, the retiring Mexican minister. The table was elaborately decorated with flowers, and large *menus* of pink and crimson satin were laid at each plate." Besides the guest of honor, the account listed Senators, Representatives, Justices of the Supreme Court, the Secretary of War, and diplomats, one of them Señor Barca, the brilliant, gambling Spanish Minister.⁵²

Before Admiral Ammen could get around to denouncing this soiree, the watchful eye of Joseph Medill had fixed it with a baleful glare, and an editorial in his paper, the Chicago *Tribune,* of March 8, headed "EADS'S ENORMOUS LOBBY," declared: "Captain Eads himself is the most audacious, unprincipled and successful lobbyist the national Capital has ever known." It railed at Eads for parties he had given and some he had not given, accusing him of financing the dinners of other hosts so as to be a "conspicuous orator" of the occasion and thereby reach the ears of certain picked influential men.

Having been goaded for years by Medill, who had "grossly assailed" his private character and professional labors, Eads came back at his tormentor with a letter which he made widely public. He painted the Chicago editor as possessed of "an innate love of misrepresentation and abuse," and of being "devoid of every sentiment of truth and honor." He was about to run out of invective when he recalled that one of the dinners which Medill had accused him of financing had been given by a Mr. Hutchins especially in honor of Washington newspapermen, so he topped off with this line: "Your attempt to disparage it makes the adage of the 'dirty bird' peculiarly applicable to you."[53]

Choice bits from both sides of this wordy battle were tossed around at social gatherings and in Congressional anterooms, but James Eads had shaken aside the whole affair. He was sparring with all his strength and skill against the promoters of rival isthmus plans, and with the stay-at-home-and-mind-our-own-business advocates. To these last he ranked as an extremely troublesome man, he was always talking about his country's share in world commerce, about expanding its trade with Asia, North Africa and South America, as though any of those remote and barbarous parts of the earth could possibly be important to the people of this enlightened, self-sufficient land. In March his bill was reported favorably in committee and Senator Vest asked to have it put on the Senate calendar. Its enemies were lying in wait for it with all the familiar devices to keep it from coming to a vote. Just the mention of it brought on an instant motion to adjourn. It was foiled and blocked. In the House it was sent back to committee. And presently the session was over.[54]

Sore over the loss of time, but certain that the growing public sentiment in favor of the ship-railway would eventually sway

Congress strongly to its side, James Eads picked up other work that awaited him. His fame as an engineer had spread over the world, his correspondence was enormous. He was a member of many different scientific and engineering societies and constantly besought for opinions. Late in the year, 1882, he sailed for Mexico, carrying with him his report on the harbors of Veracruz and Tampico. Grown to be a popular figure at the Mexican capital, he was received with the most flattering courtesies everywhere. To his surprise he found that, in gratitude for his advice on the shoaled harbors, and for his fight to link Mexico with the United States by a bond of partnership in an isthmus crossing, President Gonzalez and his Cabinet had detailed a man-o'-war to bear him back home when his mission was over.[55] It seemed strange to him to pull up at modest Port Eads in such foreign splendor, but it was part of the drama he played for the river. He lived for the day when the Mississippi would clasp hands with the Pacific by means of a ship-railway.

At New Orleans a group of citizens, devoted to the man who had so stubbornly forced the jetties upon their water road and thereby raised their port from eleventh to second place as an American export point, had arranged a social tribute to him, a lavish dinner for which "a celebrated golden service" was trotted out. Measured by the number of courses and the variety of food, it was a royal affair.

"The banquet given last night to Captain James B. Eads at the St. Charles Hotel was in every respect worthy of the citizens of New Orleans and the gentleman they sought to honor," reported the local *Times-Democrat* of December 7, 1882. "The ladies' ordinary was decorated in a highly artistic manner, the walls, gas jets and table being one long floral ornament. There were two immense maps . . . representing the jetties . . . which also contained plans of the St. Louis Bridge and the isthmus ship-railway and other of Captain Eads's more celebrated works. On the *menu* before each guest was a strikingly like photograph of the guest of the evening. . . . There was also, at each place, a card representing one of the animals, fish, flesh or fowl, to be served, bearing the inscription: 'Captain Eads, we appreciate your good work.' "

There were eulogies and reminiscences. Captain Eads had given New Orleans a rebirth, he had saved its shippers from unbearable delay losses and the tug pirates. One guest recalled paying thirty-five hundred dollars to get the steamer *Alabama* across the bar at Southwest Pass, and another said feelingly: "I have laid aground on that same bar as long as forty-five days at a time."[56]

When James Eads arose, "amid vociferous cheers," he talked of the future of the Mississippi, of a deep channel down its navigable length, of a short route to the Pacific for its commerce. His hosts, lounging back from the flower-decked tables, felt it borne in upon them that the river was their guest of honor—the slight man standing there, with the shadows from the chandeliers playing over his earnest face and rapt eyes, considered himself but its servitor. To him no ambition for the vital stream was too lofty, no labor too severe. He would plan and work for it to the last day of his life, with an army of helpers if he might, alone if he must. It seemed to each man there that he would gladly risk his career, his all, to stand beside James Eads in his battle for the Mississippi—the statistics on the port's growth had been heady, the wine lavish.

And even on many a bleak tomorrow the men of New Orleans were to lend their stanch support to the man who knew their river as no one else ever did, and loved it as few others had.

CHAPTER XVI

DEEP CHANNEL

THE upright, trim figure of James Eads, his long strong ruddy face, his dynamic voice and correct manner were familiar in the halls of Congress, in lecture rooms across the country, and at foreign ports. Portraits of him had looked from the pages of American and European journals, his projects had raised spirited debates across two hemispheres. Yet he remained something of an enigma, even to his friends. A trail of vague questioning followed him: Why did he belie his manner of a courtier by talking, even to rulers and diplomats, of his manual labors on steamboats and in diving bells? What prompted him to use the same formal but friendly approach to wharf hands and statesmen, willow cutters and kings? How could he hold enthusiastic confidence in an undertaking when it appeared to face certain failure and other men had deserted it? Why did he turn from one nearly crushing venture to another, fight singlehanded against swarming opposition, endure ridicule and abuse, sacrifice his health, imperil his fortune, alienate his friends? Above all, why did he keep on now at the ship-railway just as though Ferdinand de Lesseps were not all the while constructing a canal at Panama?

It rarely occurred to Eads, who took the utmost care to set forth an abstruse engineering problem so that anyone could grasp its fundamentals, to explain himself. He felt that he was a simple man whom anyone could understand at once. It puzzled him that polite surprise looked from the eyes of his acquaintances when he referred to his manual labors, he considered earnest work of any kind a dignified thing, worthy of pride. He treated all men alike because each, high or low, bore the intrinsic value of a human soul. He held in much the same respect the friends to whom he was always sending gifts and the enemies whom he scorched with a caustic pen. His punctilious manner—that was more to gratify his own sense of artistry than to impress others. As for his confidence in projects that seemed doomed to failure, he had often said that that was based in his reliance upon the science of engineering and a true recognition of natural law. This "faith," however, so frequently mentioned in newspaper items about him, appeared at times to transcend all physical considerations, reaching beyond his obstinate will into a realm of lighted vision.[1] His persistent undertaking of herculean tasks at the risk of health and fortune—he did that only for the Mississippi. He had easily refused to improve the harbors of Brazil, he had staked no personal wealth or exhausting effort upon the St. Johns or the Sacramento River, or at Toronto. And it was mainly for the Mississippi that he persisted in his efforts to launch the Tehuantepec ship-railway.[2]

He had never believed that Ferdinand de Lesseps would be able to complete a tide-level canal at Panama, or that he would consent to change over to a lock canal, and now the public was generally swerving to this opinion. Tales of hardship and ill success that drifted out of Panama were raising grave doubts everywhere, particularly in Washington. Earthquakes, which the optimistic de Lesseps had vowed were unknown to that part of the isthmus, rocked it on September 7, 1882, flinging a tidal wave in from the Atlantic, rending the island of Colón across, and leaving a fissure along three miles of the right bank of the Chagres River, halting the work. There were months at a time when the torrential rains drove the workmen out, washed excavated earth back into the cut and buried the machinery. Fever raged always. Dr. Walfred Nelson, a Canadian physician there, divided the year into "the wet season . . . when people die of yellow fever in four or five days" and "the dry or healthier season . . . when people die of pernicious fever in twenty-four to thirty-six hours." There had been mutiny among the laborers imported from Caribbean islands, riots, looting, arson, murder. In all, very little real work had been accomplished.[3]

With this near breakdown of the Panama enterprise, interest in the ship-railway for Tehuantepec noticeably heightened in both the United States and England. James Eads had been deeply disappointed at the conniving to keep his bill from the floor of Congress in the past two years, 1881 and 1882, and at the fact that a few men, by strategy, could edge a popular measure into outer darkness. It was especially maddening that each failure had been by the scantest margin, as if just a little more effort or alertness, or perhaps even luck, would have seen his ship-road bill through. Time was wasting, precious time to men sailing fourteen thousand miles out of their way around Cape Horn. But he would not give up. Another season, and he would scatter the small but determined opposition, as he had the jetty foes, until they gasped for breath.

Pushing boldly ahead with his plan, Eads started to Mexico in March, 1883, to commence work on the ship-railway in order to keep its Mexican charter alive. But he had an errand in New Orleans on the way: the River Improvement Commission was going to meet there and he had a touchy but vital matter to settle with his colleagues on it. The commission's work had commenced with high promise two years ago. More than three hundred workboats including floating pile drivers that operated by hydraulic jets, a mattress barge carrying a steam loom which turned out a continuous mattress woven of wire and brush fed to it, floating machine shops and boardinghouses, and many flats of stone and coal were assembled.[4] The chugging, digging, weaving and shouting raised more uproar than the river above the delta had ever known in peacetime. Scores of miles of jetties were swiftly built at the troubled sections that had been selected, a timely flood filled the woven mattresses with silt and scoured the bed. In a short while the once-pitiful five-foot channel in the Plum Point stretch had been deepened to twelve feet, and the six-foot depth of the Lake Providence reach had become fifteen feet—altogether a phenomenal length of excellent navigation channel had been produced. James Eads thought that the success of his method had been proved, but while he was in England lately on a business trip he heard that his jetty bank contraction had been abandoned by his fellow members, who had decided, too, that broken levee banks need not be mended, as the escape of water through them might be providential when the river was too full.[5] Eads had left England at once for the United States, his wrath gathering all the way.

It had burst forth as soon as he landed. The costly bank revetment, which ignored the narrowing of overwide sections of the stream, would not give the river increased room for floods or navigation channels an added inch of depth. In fact, the channels had been left completely out of consideration, Eads accused. After his tirade, the contraction method was mildly restored, but presently another point of difference had risen sharply between James Eads and the others of the commission. The Atchafalaya River, flowing down through eastern Louisiana, had lately had its driftwood raft removed from it by the state, and thereafter, with so much new room in its bed to fill, it reached up and tapped Red River floods, enticing more and more of that stream's high water away from its former journey to the Mississippi. This, too, was considered fortunate by the majority of the commission, for it would lighten the flood overflow on Mississippi lands below —they even thought that it might be well to increase this defection of Red River, and decided to lay a sill across its mouth where it flowed into the Mississippi, thus virtually divorcing the two streams. James Eads was horrified at the idea, and determined that the thieving Atchafalaya should be kept from filching any water that rightfully belonged to the great trunk stream.[6]

Now, as Eads was ready to start to Mexico, the other members of the commission were on a leisurely tour of inspection down the river, aboard the government-provided steamboat, the *Mississippi*. Eads hurried ahead of it, likely in order to get in a visit to South Pass. On a warm March evening James Eads drove to the foot of Lafayette Street in New Orleans, where the government boat had moored, and went aboard.

There was a politely screened tension as the commissioners sat down on deck. A fitful breeze played hide-and-seek, the lights of the town spread away like a jeweled carpet in the evening haze. The discussion moved along placidly enough until it reached the topic of the Red River sill. James Eads stated his objections. They were received coolly. He peppered the scheme with rapid-fire arguments, but they were like buckshot rattling against a wall. The other members held firmly that the Red River floods must not add to the Mississippi overflow, nor the Mississippi floods back up into Red River. The sill would be laid.

What about navigation? Eads demanded. Was it of no importance? If deep boat channels were provided, they would do more than anything to prevent overflow. Look at that line of ships, steamer chimneys and schooner masts rising like a curving shore forest against the night sky as far as one could see, ships from everywhere, the largest vessels on the globe, come up through South Pass. Look at the long double line of steamboats above the city, fetching and hauling for the broad valley. Was all that traffic not worth considering? Perhaps the other gentlemen present did not realize what nearly superhuman effort it had cost to provide an adequate flow of water into South Pass for a decent shipway—if Red River ceased to supply a substantial part of that water, New Orleans, lying yonder, so opulent and contented, "would find itself left on the banks of a dead lagoon."[7] Red River must not be shut off from the Mississippi.

Right or wrong—and later engineers have both agreed and disagreed with him—James Eads was stormy in this contention. But when the meeting closed on a high pitch he was standing alone, the others arrayed solidly against him. The cursed sill would be built, he feared, nothing that he could say would prevent it. The scene had been painful and futile. He stalked from the boat. Convinced that singly he could do no good on the commission and that his very presence on it might be harmful, seeming to lend acquiescence in the plan he deemed so destructive, he stopped in a telegraph office and sent his resignation to Washington.[8]

Reports leaking out of the meeting brought on a prodigal use of black ink in news headlines: "Captain Eads Very Angry," printed the New York *Times*. Captain Eads had walked out on the River Commission . . . Eads had abandoned the Mississippi to its fate . . . Eads had turned his back on the big river, other papers hinted. Left the river to its fate, echoed once-hopeful ports along it. But James Eads had no idea of tossing away his dream of a deep Mississippi channel, he would merely bide his time until the present commission came to an end, as commissions always did, then find a way to carry through his ambitious plans. Until then he would throw his whole energy into furnishing a Pacific outlet for its commerce.

As was his habit, he dropped from his mind the disappointment of yesterday and gave himself fully to the next work at hand. He was on his way, despite the lack of an American charter, to break ground on the ship-railway, according to the terms of his Mexican charter that work must start within two years of its granting. And he was sure that after another session of the United States Congress he would carry on the work in volume, making up some of the time that had been lost. In Mexico City the little drama was arranged, and on April 23, 1883, a small gang of peons began clearing the way for the actual ship-road. Three days later Mexican officials held an inspection and a formal opening of the Tehuantepec ship-railway, presenting James Eads with a certificate that the work "was begun *bona fide,* within the two years required by the concession."[9]

One cannot tell now how sincere or hollow all this may have rung at the moment, but it served to bring the project down from a mirage on the horizon to the earthiness of creaking saws and bending backs. It pinned upon it many eyes that were too myopic for distant chimeras, and it bound by ceremony the cordial friendship that had grown up between the American engineer and the Mexican government. The ship-railway would be built, James Eads felt with all his heart, and he communicated this confidence to President Gonzalez and ex-President Diaz, both of whom wanted avidly to believe that an isthmus crossing, with its gift of golden prosperity, would happen to their country.

Back in the United States, James Eads paused at Compton Hill.

The trees were lacy with new green, the air full of the song and twitter of birds. The children of his stepdaughter Josephine, who, with her husband, Estill McHenry, lived there with lonely Eunice, and the children of Jacob Stiel, the gardener, followed him over the grounds and garden. He went to see his daughters, Eliza and Mattie, who lived down Lafayette Avenue a piece, side by side in houses he had given them. He distributed gifts among their children and the grandchildren of Sue Dillon Stevens, as usual.[10] He tried to arrange to have Eunice with him more, but she was not fond of travel—at least, not at the pace he kept up. He saw some of his friends, talked about the river and the ship-railway, and left for the East.

In New Jersey he organized the Eads Concession Company and assigned to it his Mexican charter rights.[11] The way would be ready for a rapid pushing of the ship-railway construction the very instant that Congress approved it. He expected to sail for England soon, but it appears to have been some months later when he settled down for a short stay in the Royal Hotel, Blackfriars, London.

Part of his business there was having a working model of the ship-railway made. If people could see a miniature shiproad at work, tiny caissons lifting a vessel, locomotives hauling it on tracks, turntables wheeling the car about, they could not help realizing how practicable it was.

When the "large and superb model" was completed he displayed it in London where many eminent engineers and shipowners came to stand in the crowds before it. Thirty feet long in all, taking up the whole side of a large room, the model consisted of a basin, a lifting dock and accessories, a section of track with six rails, a floating turntable, two perfect little locomotives, and "an elegant steamship" six feet long. The diminutive lifting dock, carrying the cradle and loaded ship, was raised from a basin of water by busy pumps, small hydraulic governors steadying it. The locomotives, puffing smoke behind them, drew the burdened cradle along the tracks, the turntables swung them about, the locomotives hauled the cradle and all back to the starting place, the caisson descended with it in the basin and the ship was floated off. Eads himself was on hand to answer the pelting questions, and gained applauding comment, even from the hitherto most skeptical.[12] It looked as though the amazing model, going about its pygmy business so smoothly, would convert all who saw it to the ship-railway project. Pictures of the model appeared in scientific journals, excited speculation about the Tehuantepec venture hung in the air over many countries.

It was midwinter of 1883-84 when James Eads arrived in Washington with his model. He had two interests there: the endorsement of the ship-railway plan and the adoption of a workable, over-all design for the improvement of the Mississippi. They had to go hand in hand. Congress had lost faith in the creation of a deep river channel since his jetty scheme for it had been discarded, appropriations had failed, and the middle country was seething in angry protest, erupting like bubbles in boiling mush. A conference of river interests that had been called in St. Louis had urged that a special River Convention be held at the National Capital[13] and a concerted demand for a deep waterway be made within earshot of Congress. They would toss the whole stale mess of negligence and broken pledges into the air—and let the legislators try to dodge!

Washington, wintry cold, was, to James Eads, principally a familiar battleground. He could look about him and see how lobbyists and legislators were deploying their forces for the various struggles. To Eunice, stately and beautiful, her now white hair piled high on her head, a visit here was just another series of social triumphs and thrills. The "season," at its height, shifted its kaleidoscopic bits, grim or merry, cheap, dignified or fantastic, into varicolored patterns. Social chatter breezed about in little flurries: who sat in whose theater box to see Henry Irving and Ellen Terry, Joe Jefferson or J. K. Emmet . . . who had been on what excursion to the recently completed Washington Monument . . . the wealthy Leiters of Chicago had rented the Blaine mansion . . . Minister Barca from Spain had killed himself over his gambling debts . . . Joaquin Miller, the poet, garbed like a Sierra gold digger, was entertaining in the log cabin that he had daringly put up right among the conventional' mansions on Meridian Hill.[14] Invitations flew about and everyone watched to see what everyone else accepted or rejected, trying to predict, thereby, the destiny of bills on which hung the hopes of ranchmen in the Southwest, fishermen in New England, Irish immigrants or Chinese coolies. The game of prognostication by social signs was diverting or deadly serious, but everyone played it.

In February of 1884 the River Convention elbowed into the crowded Capital. James Eads, although he had resigned from the River Improvement Commission, longed to see the convention accomplish something substantial for the Mississippi, but there seemed scant hope of that. The delegates were jealous and touchy, they had come brimming with specific demands and eloquence, but empty of unity. Each river stretch, each port and region, clamored for especial consideration. The meetings were tense, the speeches pugnacious. The whole affair appeared about to flare into a melee of sectional strife when a Mr. Turner, of Tennessee, took the floor. James Eads must have smiled to see the delegates, stubbornly braced against whatever particular plea was in store, relax in wonderment at a quietly told Buddhist tale from his own favorite poet, Edwin Arnold:

Once, long ago, Turner related, when famine had settled over a district of India, a starving lioness and her cubs crept about the parched jungles and fields looking vainly for food. There happened to pass along not far from them a man who was so spiritually enlightened that he felt that the interests of all creatures were one with his own. Spying the beasts, he slipped behind some undergrowth and watched them. Their gaunt bodies and desperate eyes filled him with pity. "He who stood in the bushes," Turner raised his voice dramatically, "was a living child of God's Eternal Brotherhood. . . . At last, knowing no animal, knowing no East, knowing no West, and looking upon all alike, he hurled himself from the bushes and cried, 'By the Eternal Brotherhood that should live in the human heart, feed your starvelings, O Lioness, upon my body, and let God take care of my soul.' "[15]

There was a gust of applause, but when it died down no one had been converted to the exalted length of wanting a competitor to fatten upon his sacrificed carcass. However, sectional strife, made self-conscious, was less noisy through the rest of the session. Still it thrust an aggressive shoulder into every proposed compromise, and the petition sent to Congress, at the end of the bickering, was so lacking in force that it could be easily pushed aside.

The valley was not yet ready to treat its vast river as a whole, James Eads regretted. Perhaps only some striking common interest would reveal to the ports and sections that they were interdependent, that a prosperous Cincinnati or Minneapolis meant a prosperous St. Louis or New Orleans. The jetties had stimulated the various, but selfish, demands for channel improvement here and there along the river, and it might be that an isthmus crossing to the Pacific would weld the competitive pleas into a single powerful appeal.

The ship-railway bill must pass without further loss of time, Eads resolved. Carefully, determinedly, he organized his forces, he talked to legislators, rallied his friends about him. He set the ship-railway model up in a room of the Capitol, watched Washington flutter away from its ordinary social rounds to exclaim over the miniature locomotives and caissons going through their routine. He beamed at the excited buzzing about it later over teacups and wineglasses. He heard the wails of the Nicaragua route advocates when they failed to obtain from Congress a renewal of their expired treaty with the Central American country, and their accusation that his ship-railway model had defeated their attempt. He was encouraged by comments of many statesmen that the model had clearly shown that a route eight hundred miles nearer the United States than Nicaragua and twelve hundred nearer than Panama was "entirely practicable and superior." He put forth every effort, fenced alertly with all opposition, but when the bill was read in the Senate, it was referred back to the Committee on Commerce, and there it was wrangled over for months, never reaching a Congressional vote.[16]

CHAPTER XVII

THE LONG FIGHT

COMPTON HILL had never looked more beautiful to James Eads than in the last months of 1883. The familiar slope with its tree-bearded cheek turned to the morning sun, a deformed pine fleeing with outstretched arms from a regiment of oaks, the fox squirrels that scolded each other over their winter store of nuts, the gardens, the greenhouse, the long drawing room, the quiet library—he would soon be leaving all this, leaving St. Louis with its tree-lined brick streets, the waterfront and its border of steamboats, the river. It had been fifty years since he landed, "barefoot and coatless," about where the West Abutment approach reached inland. Much had changed since then, but not the river. It wound out of the gray north and hurried past to the misty south, murmuring to itself just as it had on that morning so long ago when it picked up the copper of flames and shadow of smoke from the burning boat to toss them aside. The river, dancing along with its rich traffic to the Gulf—it was so that he might be better able to give it a Pacific outlet, a world importance, that he was planning to move away. To New York. From there he could reach Washington in a few hours if necessary, and he could keep in close touch with a large group of Pittsburgh men, headed by George Westinghouse, George Whitney and William Thaw, and another group in New York, prominent among them C. J. Ryan and R. M. Van Arsdale, who were ready to support the ship-railway to their utmost. His going would be made easier for him because Eunice agreed to it so readily—she would have him with her more, and she enjoyed the big

THE LONG FIGHT 271

city and its shifting pageant. Josephine and her husband, Estill McHenry, would stay on at Compton Hill. Someday he and Eunice might come back there.

It was in the spring of 1884 that the uprooting and packing began. It must have seemed almost a desecration to James Eads, but he was spared from seeing much of it. He had been summoned to England on a special mission—to give an opinion on a projected ship canal from Manchester into the Mersey River estuary. This canal was to be thirty-five miles long, as deep as the Suez Canal then was, and much wider. It was a tremendous enterprise, and the subject of a long harsh controversy, opposed and defended in a series of legal contests. Now the British Parliament was inquiring into it, and the Mersey Docks and Harbor Board of Liverpool, who contended that the canal would eventually damage the depth of their channel over the bar at Liverpool, in seeking the highest possible authority on such matters had agreed to call in the noted American engineer.[1] Eads had sailed at once for England.

There he made an intensive three weeks' study of the Mersey estuary, then appeared before a committee of the House of Lords to give his opinion. The canal as designed, he told them, extending several miles into the Mersey, would certainly lead to the destruction of the Liverpool bar channel, ruining the harbor. His argument, with the illustrations drawn from notes and maps some of which were a century old, were so conclusive that the committee condemned the long-debated design, and later Parliament discarded it. A very different plan was finally used, the canal running from Manchester around the south side of the Mersey to the Estham locks. The grateful Docks and Harbor Board paid Eads thirty-five hundred pounds, the largest fee given an engineer, up to that time, for three weeks of effort. Scientific journals commented widely upon his argument before the House of Lords committee, appraising it as "one of the best engineering papers extant."[2]

Eads had hardly returned home when he received a notice from England that the Albert Medal, instituted two decades before as a memorial to the Prince Consort, had been unanimously awarded him by the Society for the Encouragement of Art, Manufactures and Commerce, with the approval of his Royal Highness, the Prince of Wales, president of the society. The notice invited Mr. Eads to receive the medal personally from the Prince's hands. It would be quite an occasion, for the tribute had never before been awarded to one of American birth.[3] And so there was a hasty voyage back to Britain almost before he had recovered his land legs.

It was early on a mid-July evening, the sun well up in the sky, when James Eads alighted from a carriage before Marlborough House, the London residence of the Prince of Wales, where the ceremony of award would take place. It was a starched and breathless moment, with the full council of the society, men of scientific profundity and note, looking on as the stout, florid Prince Albert Edward presented the medal to the slender ruddy engineer. Although the honor was likely bestowed as a recognition of the timely advice that would save the vital harbor of Liverpool, the discreet resolutions of award read, in part:

"Mr. James Buchanan Eads, the distinguished American engineer whose works have been of such great service in improving the water communications of North America, and have thereby rendered valuable aid to the commerce of the world."

His Royal Highness made some formal, flattering remarks. In his reply Eads said that he "recognised the award as a token of good will towards his countrymen, rather than for any merit of his own; and the fact would be received by them as another of the many evidences of sincere solicitude for the happiness and prosperity of the United States which had been so often shown by her Most Gracious Majesty the Queen, her Government, and her people . . . which had assured his fellow citizens that their kin across the sea were their truest friends, and that the homely adage 'blood is thicker than water,' had lost none of its force because they were separated by an ocean from the land of their ancestors."[4]

By the time James Eads had returned from England, Eunice had the New York home, which he had bought in her name, put to rights. It was a handsome Victorian red-brick house, lavishly trimmed in brownstone. An impressive solid brownstone banister wall rose up in front of steps that ran parallel to the house, pausing on a landing before sweeping on to a spacious entrance balcony. The place was so far uptown—at 40 West Fifty-third Street—as to be almost suburban. The glass bay that stretched nearly the width of the house at the second floor looked toward Central Park, a few blocks away.[5] The park, decorous enough for a short distance, soon became a restful wilderness flanked on either side by rugged vacant land on whose lofty outcrops of granite perched the shanties of immigrant squatters, reached by ladders. Sagging roofs, awry makeshift chimneys, the colorful family washing forever strung on haphazard lines, and the summit-minded goats made up the silhouetted sky line. A series of truck farms stretched on lower ground between small settlements that dotted the way to the distant village of Harlem.[6]

Not far from this new home ran the fast and popular elevated trains, their steam engines dropping live ashes on dodging pedestrians and maddened horses. To James Eads these were unbeautiful, noisy things, but efficient. On mornings when he was not minded to drive his carriage down to his distant office at 34 Nassau Street, he could ride the elevated along with other frock-coated and silk-hatted elderly gentlemen of big business, as well as with the more youthful wearers of informal sack coats and derby hats or even of uncouth jackets and caps.

Around Nassau Street all was rush and confusion. Wagons and carriages mingled in a steady stream, pedestrians made terrified dashes from curb to curb, an occasional high-wheeled bicycle was guided through the tangle, its owner afoot. The ever-present hand organs kept up a medley of wails above the clatter of hoofs. Far beyond, in the bay, the masts of many sailing ships rose above steamship chimneys. Between uptown and downtown lay a conglomerate region splashed with such glamorous isles as the theater district of the Bowery and the shopping center of Fourteenth Street where ladies in tight-bodiced dresses with bustled polonaises tripped in and out of Macy's Drygoods Store.[7]

James Eads already had a tidy circle of friends in New York, and Eunice, ever in her element as a queenly, efficient hostess, was soon moving easily in society circles that had presented glazed, dentless rims to less sophisticated newcomers from the West. Eads's most cherished friend here was General Grant, who lived not far away, at 3 East Sixty-sixth Street. The General was not well and did little visiting, but he could often be found walking in Central Park, or standing before the cage of "Mr. Crowley," a chimpanzee whose antics helped to fill an occasional blank space in newsprint.[8] The friendship of Grant and Eads had weathered many a crisis since destiny brought them together to reclaim the lower Mississippi and its border states for the Union.

There was an intermittent stream of out-of-town guests at the brownstone-trimmed house, lawmakers from Washington, industrialists from Pittsburgh, and friends or relatives from St. Louis. There was earnest talk about a deep Mississippi channel and the ship-railway, but there was time for nonsense, too, and for poetry. A young granddaughter of Cousin Eliza Dillon's, who had paused here on her way to Europe, had to listen willy-nilly while her Cousin James read *The Light of Asia* to her—indeed, he could recite much of it from memory. The Oneness of the universe, of suns, planets, men and beasts, flowers and stones, all of them part of the Almighty Being, breathed out on the Eternal Breath, held him enthralled. He made the lines subtle and tender, ringing and masterful. Whether his young guest was swept off her feet or not, she was a politely rapt audience, and as she boarded the steamship for her voyage, Eads handed her a hundred dollars for having been "such a good listener."[9]

But the master of the house was as often missing when statesmen, engineers, fashionables, kinfolk, or persons in distress climbed the steps to his door as he had been when his St. Louis friends or visiting celebrities drove out past Lafayette Park to Compton Hill. Even though, in 1885, early in his New York days, he declined a request by the government of Portugal to deepen the harbor at Oporto,[10] and a second plea from Dom Pedro II of Brazil, baited with a promised fee of twenty-five thousand dollars, to improve a river harbor in Brazil,[11] his work called him afar and often. His travels were all but endless.

Among the lesser interests of James Eads while the ship-railway held the center of his mind was a bridge he had planned, at the request of the Grand Vizier of the Sultan of Turkey, to span the Bosporus—a change in grand viziers prevented its building[12]—and two jetty systems that he did not construct. One of these was the Jacksonville, Florida, project that he had recommended some years before. After his visit to Jacksonville, and his advice, that port, given new hope by him, had sought aid from Congress to build jetties at the mouth of the St. Johns River. The War Department sent there General Q. A. Gillmore, of the Engineers Corps, who made new surveys and designed submerged jetties instead of the abovewater walls that Eads had said were necessary. Spasmodically through the years, as funds could be obtained, the work was carried on.[13] It was still far from completed—which made slight difference, James Eads thought, for the submerged walls would never be able to do much for the troubled channel.

Of more poignant concern and perplexity to him were some jetties at Galveston, Texas, which were yet to gather about him one of the most whimsical squalls in which he had ever been caught. These jetties, begun soon after the work at South Pass was commenced in 1875, were, in James Eads's opinion, the most amazing and amusing harbor-improvement effort in existence. For nine years, during the four years that the Mississippi jetties were building and five years since then, he had watched and wondered at this Galveston project. And two years ago, in 1883, the project was clamoring on his doorstep for him to take it over.[14]

Galveston Island, a low strip of sand nearly thirty-five miles long and two miles wide, lay parallel to the Texas coast at the entrance to a sheltered bay. It was to a corner of this island that Jean Lafitte, the glib and gorgeous, having played out his notorious dual role of pirate and patriot in Louisiana, brought, in 1817, his ships and rabble horde to dart out and prey upon passing coastwise vessels. And from this colony, called by him Campeche, he was routed by an American gunboat four years later, and the island was left nearly forgotten.[15] But when Texas became a part of the United States, Galveston Bay, providing the finest natural harbor between the Mississippi and the Rio Grande, became important. A flood of products from the rapidly settling plains began to converge upon it, the ships of many countries unloaded at its flat beach, and complaints of a sand bar that sea currents had stretched outside its entrance reached Washington now and then. In 1873, soon after the jetty plans for the Mississippi mouth were published, Major Howell, then dredging at Southwest Pass, designed jetties to deepen the channel over the Galveston bar. It was this design, with the jetties to begin far out from shore and run underwater, their tops showing only at low tide, that caught the quizzical attention of James Eads—it was the construction and its results that held his surprised interest.[16]

The two jetty projects, several hundred miles from each other, the one at South Pass financed by Eads and his friends, the one at Galveston financed by the federal government, with no risk to the builders, had gone on season after season. By July of 1879, when the South Pass jetties had deepened the channel over the Gulf bar to thirty feet, an improvement of twenty-two feet over the original depth, the Galveston jetties had produced no depth change whatever. By 1880, the north jetty at Galveston, which was constructed of gabions, or long bricks formed of hollow willow frames filled with sand and plastered on the outside with cement, had reached nearly two miles to sea. Still there was no deepening of the channel. During this year Major Howell was replaced by Colonel Mansfield, who, when he took charge of the works, began to look around for the expensive north jetty, and finally discovered it sunken in the sand bottom of the bay, much of its top lying level with the bay bed. He rebuilt part of it, then turned to the task of providing a south jetty. In 1883, after three years of effort, Mansfield reported that the new jetty, the south one, more than four miles long, was completed, having been built up to low tide. He was puzzled, though, he admitted, because the scour it was intended to produce took place *outside* of it, while sand that the waves carried over its top shoaled the channel more and more.[17] Galveston was growing nervous. The people had no faith in the attempt that was being made to improve the harbor, and appeals to Captain Eads to take over the project had been refused—he would not encroach upon work that was in the hands of government engineers.

Some months after the completion of the south jetty, Congressman C. R. Breckenridge, a member of the River and Harbors Commission, journeyed from Washington to see the walls that had been building in Galveston Bay for so many years. Colonel Mansfield took him out in a boat at low tide to inspect the oddly behaving new jetty, but, to his dismay, it was nowhere to be found. After hours of punching down in the water with a long pole the recently finished south jetty was discovered, already sunken seven or eight feet. As for the north jetty built years before and partly restored, what, demanded Congressman Breckenridge, had become of it? Colonel Mansfield and his aides explained that they had not yet located the north jetty. This could meant that they had never decided just where to construct it, or it could mean that a built jetty had long ago been utterly lost in the sands of the bay bottom. The report that Breckenridge made of his findings sputtered with indignation.

Sourly amused over the unhappy project, James Eads related some of the incidents to his friends, and they made good telling: An anxious congressman, in tow of a perplexed government engineer, hunting, "with blistered hands and a sad heart," the elusive new south jetty by means of a ten-foot pole! And, later, "four of the oldest, ablest and most astonished officers of the United States Engineers sitting in earnest consultation over the incorrigible and moribund north jetty that was never located!"

The fruitless tinkering in the harbor had finally brought the patience of the Galveston people to an end. While all but the smaller Atlantic ships were closed out of their port, the largest vessels afloat were sailing through the South Pass jetties where the minimum depth was now thirty-three feet. If Captain Eads would take over the channel improvement, all would be different. Representative men from the city's business and political life held a meeting in December of 1883, and voted to send another plea to the builder of the Mississippi jetties.

The appeal reached Eads in England. He did not want to shoulder this heavy undertaking now, he had been ill and was awaiting strength to resume his fight for the ship-railway. But the Texas port was in a critical dilemma, it had suffered nearly nine years of baffled hope while the congestion of products from the vast area it served had doubled. He replied that he would consider the offer.[18]

The government engineers, who had striven, with inadequate funds and scant experience in large hydraulic problems to create a deep channel over the Galveston bar, were piqued. Defending their ineffective works, they explained that their jetties, so far, were only "tentative." They argued that an engineer should be cautious and build a jetty here and one there so as to find out how they worked, because "it was quite possible that one or the other might require removal."[19]

In a later spicy review of this defense, James Eads observed that if the government should put an Army surgeon in charge of harbor work he would, of course, proceed on such unfamiliar ground with due caution. "I do not believe, however, that the surgeon would, when called upon to amputate a man's leg, adopt the tentative system . . . and cut off a little today, and a little more tomorrow, prepared to change his plans if he found that he had got hold of the wrong leg!"[20]

Coming now to a situation that had annoyed him a long time, he wrote with some indignation: "In the whole history of the government in its dealings with officers of the Army, there is nothing in such prominent contrast as its treatment of the Engineers' Corps, and that of its other corps. Mistakes or lack of judgment on the field in the heat of the contest . . . have been rewarded with courts-martial, loss of pay, of rank, or disgrace;

whilst notwithstanding absolutely inexcusable blunders made by several officers of the Engineers' Corps after abundant time for investigation and involving immense loss of public money and vexatious delays to great and needed improvements, they have been rewarded with promotions."[21]

As soon as he could get around to it without sacrificing the prospects of the ship-railway, James Eads, now settled in his New York home, made an exhaustive study of the Galveston bar problem. Toward the end of 1884 he announced that he could design and build jetties that would furnish a thirty-foot deep channel over the bar at a cost of $7,250,000, as close a sum as he could figure—the walls would need to be longer and stronger than those at South Pass, the material more expensive. The Texans themselves would have to put the necessary legislation through Congress, he stipulated, for he had not forgotten the humiliation he had endured before he succeeded in getting the Mississippi jetties authorized, and he wanted to spare himself another such experience.[22]

At once all was eager expectation at Galveston. The Texas legislators and most of the state's press endorsed Eads's proposal. A bill embodying it was introduced in Congress in January, 1885. Purposely keeping afar from Washington, James Eads made another trip to Mexico.

Perhaps if he had been at hand, Capitol Hill would not have rung so with his name. The Army engineers, who had not yet found the lost north jetty at Galveston and were in doubt at times as to the whereabouts of the new south jetty, came forth with an accusation that Eads was merely rigging up a villainous scheme to extort a large sum of money from the public treasury.[23] They found sympathetic legislators to repeat their arguments. Speeches attacking James Eads and his proposal resounded in both houses. Flushed defenders of the absent engineer demanded recognition. "I am no more a hero-worshiper than any other gentleman on this floor," Congressman Newton C. Blanchard said earnestly as he rose in the House of Representatives on January 31, "but I am here to do justice to James B. Eads."[24] And this he did, recalling the maligned man's impeccable honesty, his monumental works, and the esteem in which he was held abroad. The debate was in full blast three days later when the Hon. W. P. Hepburn climaxed his attack upon the Galveston proposal with this warning: "It puts the entire Treasury of this Government into the hands of this man Eads!" He advocated that Eads be set aside as designer or contractor and, instead, be hired at a salary of thirty-five hundred dollars a year to carry on the work.[25] A clause to this effect fastened upon the bill and still clung to it like a bur when the whole matter was dropped into an omnibus bill crowded with diverse river and harbor projects. The press made much of all this.

And so it was that James Eads, in Mexico, met himself in sensational headlines in the American newspapers: The Objectionable Scheme . . . The Questionable Plan of James B. Eads. He was stung by the item that he was to be hired on a salary as though he could not be trusted with public construction funds. And he suspected that it was intended to put him under the orders of the very men who had bungled the Galveston work for years. He sailed for New Orleans. From there he blazed back at some of the newspapers, and hurried on to Washington. Here he demanded to be heard by committees of both houses of Congress, insisting that his principal detractors, General Newton and Colonel Mansfield, be present.

By the time he stood before the legislators he had somewhat leashed his anger. Cutting short his countercharges and denials, he coolly and patiently explained why the lost jetties at Galveston had not worked, and what must be done to secure a deep channel. It had a damning effect upon the case that had been built up against him. He pointed out that his proposal to build the Galveston jetties was an open bid. Anyone was privileged to underbid him, and some "parties" had considered doing so. But they had sent Mr. O. Chanute, a well-known engineer, to look over the site and he had reported that there was no margin of profit even at the figure that had already been named.[26]

Having set forth his side of the case, James Eads withdrew his proposal. Galveston was stunned. The businessmen held excited meetings. The City Council went into session and voted to

THE LONG FIGHT   281

offer Captain Eads a contingent fee of a hundred and sixty thousand dollars above the pitiful salary that had been mentioned in the harbor bill. They sent their offer hopefully, but James Eads refused it. He could not very well consent to take it and work under the control of hostile engineers, carrying out plans that he believed would end in failure.[27] Presently the omnibus bill bearing what was left of Galveston's hopes careened to defeat.

There were reproachful mutterings all over Texas directed at the two Army engineers who had so valiantly driven Captain Eads from the field. If these gentlemen were at all daunted they were able to save their pride by coming forward with a ringing announcement. Some time before this, phenomenally, at the end of ten years of effort, the depth over the Galveston bar had suddenly increased from the original twelve and three-quarters feet to thirteen feet—an improvement of three *inches!* Certainly no one had any reason to pronounce their jetties a failure. At the same time they admitted in some bewilderment that, for reasons totally obscure to them, there had occurred an additional depth of four to eight feet, not inches, on the outside, the *wrong* side, of the south jetty.[28]

Galveston was disgusted with the report, Washington was stupefied. James Eads pondered it with amusement, and regret. The Southwest was a rousing giant of production, it needed a deep seaway at the excellent Texas harbor. No doubt this would someday be provided, for Galveston was still prodding Congress about it.[29] As for himself, he was out of it. He would devote his whole energy now to his ship-railway plan. His model, the steamship of which he had named the *Carmen* for the beautiful young wife of General Diaz, again President of Mexico,[30] had been taken about the country by Elmer Corthell to illustrate his lectures in behalf of the project, and it would later be set up as an attraction at the New Orleans World's Fair. Now it was on view in a basement room of the Mutual Life Insurance Company of New York, at Nassau and Liberty streets,[31] and out-of-town guests at the Eads home near Central Park were sure to jaunt by carriage, perhaps more than once, the long trip down to the end of Manhattan to see it.

CHAPTER XVIII

THE TIRED WARRIOR

ALONG the valley rivers men had come to accept as a certainty that the Mississippi, by one means or another, would soon have an opening to the Pacific. Now they realized that they could no longer lean back against this rosy hope. The work at Panama had fumbled along for five years, a hundred and fifty million dollars had been spent upon it, and only one-fifth of the excavation had been made.[1] The recent sanguine announcement by Viscount de Lesseps that the canal would be completed within the promised time was, to put it mildly, unfounded. In any case, Panama was too far away. The valley should lend a stronger hand to the Tehuantepec ship-railway.

In St. Louis, which had been caught up mildly in a discussion of whether an injection for rabies was a depraved attempt to turn a human being into a dog, the determination to put through James Eads's Tehuantepec project had come down to points. A meeting of fifty prominent citizens was called in November, 1885, in the gentlemen's parlor of the Southern Hotel. The oratory was piqued and accusing: Why could not the United States meet the generosity of its neighbor Mexico? Congress, which had handed outright to the transcontinental railroads nearly three hundred million dollars in bonds and land, could well make a loan of one-sixth that amount for the benefit of water commerce.

"A careful examination of the working model of the ship-railway would convince any intelligent mind that the great engineer, Captain Eads, has successfully solved the problem of our time," thundered ex-Governor Stanard of Missouri, and closed his speech in a telling flourish: "But when the eagle of American genius, with her broad pinions resting, as it were, upon the clouds, shall reach down her talons and seize the great ship with its burden of rich merchandise, lift it out of the sea and transport it unharmed from one ocean to another, then, indeed, may the fabled god of water have occasion to frown!"[2]

General Sherman was in the audience. Called to the speaker's desk, he said that it was his belief that a ship would be safer on the ship-road than when buffeted by storms as it rounded Cape Horn. As to whether a large ship could be lifted from the sea to a land track, he quoted Aristotle, who had declared that, given a fulcrum, man could raise the world. The meeting, now at high pitch, passed resolutions endorsing the ship-railway and praying Congress for aid in its construction.[3]

James Eads came back to Washington in December, 1885. Surely, he thought, statesmen knew by now that it would be fatuous to wait upon the tragic effort at Panama to furnish an isthmus-way. Reports of travelers painted the blundering and mismanagement there as appalling. Nothing like it had ever been witnessed. The number of bureaucratic hirelings was preposterous, the recklessness with funds almost incredible. A director-general lived in a hundred thousand dollar mansion in winter and a costlier one in summer. Another official had a forty thousand dollar bathhouse, and still another a fifteen thousand dollar pigeon cote. Extensive landscaped grounds were the rage, personal servants had already cost a substantial fortune.[4]

"In all the world," wrote James Anthony Froude, "there is not, perhaps, now concentrated in any spot so much swindling and villainy, so much foul disease, such a dungheap of moral and physical abomination." Along with many other items of waste he mentioned that "half buried in mud lie about the wrecks of costly machinery consumed by rust, sent out under lavish orders and found unfit for the work for which they were intended." Only the excellent hospitals lent an encouraging note.[5]

But Washington was too preoccupied to take in this far-off scene. The town had never been so full of diversion and gossip.

Phonographs with earpieces fascinated curious crowds wherever they were displayed, rivaling the ship-railway model. Society was gathering for its most brilliant season. Young Theodore Roosevelt, lately defeated candidate for mayor of New York City, and his bride had taken a modest house in the Capital; Senator Leland Stanford had moved into a big house on K Street—Mrs. Stanford "wore diamonds like a coat of mail."[6] Up on Capitol Hill there was talk about the extension of civil service reform, about suspending the compulsory coinage of silver, about levying a tax on oleomargarine. Across this James Eads lanced his persistent theme: commerce should not be hampered by a ridge of land between the great oceans while man's ingenuity could devise a way across it. Some paused to listen to him when he stressed the fact that Mexico had added to its already magnanimous grant another million seven hundred thousand acres of public land and an offer to guarantee one-third of the ship-railway's revenue for fifteen years after the road opened.[7] They had been told this before, but had given slight heed to it. It left little enough for the United States to do to provide an isthmus-way.

Christmas passed and the year 1886 slipped in almost unnoticed as friends and opponents of the ship-railway began to line up for the hardest battle of the long isthmian war. Countless words—and some of them *such* words, grandiloquent or abusive —were flung on paper or into the frosty winter air. Elmer Corthell took time from his lectures over the country to plead the case of the Tehuantepec ship-railway before a Congressional committee, waxing dramatic with the familiar cry: "A route to India, to Cathay!" The United States should not remain cut off from the riches of the Orient! "The fleet of Solomon," he recalled, "in one voyage brought back from the East gold valued at fourteen million dollars. In the days of Augustus Caesar the products obtained in India in excess of those exchanged there amounted to $40,000,000. . . . Of the vast trade of the Pacific we enjoy but four per cent, and yet these countries would be the best market for our manufactures if we could reach them economically. . . . A ship-railway at Tehuantepec would save in cost of transportation about twelve thousand dollars per cargo over the present route around Cape Horn."[8]

Opposition to the project was as stiff as ever, with Admiral Ammen leading it. His canal company had had to forfeit its concession from Nicaragua for want of encouragement from the Congress or people of the United States, and this he laid at the door of James Eads. Lately Ammen had become highly incensed over a reference Corthell had made, in a public address, to the "constructive genius of James Eads," and had at once put out a pamphlet of objections to the ship-railway and its projector, dwelling especially upon the danger of a ship's being wrecked on its overland trip by gales or earthquakes "which seemed to be a probable monthly, if not weekly, occurrence." But, he surmised witheringly, wrecks were doubtless very interesting to Captain Eads. That ever-busy, never-ceasing, unscrupulous meddler had become a financier and helped wreck the State Bank of Missouri; he had built a bridge which, although it had not fallen down *yet*, surely would have except that its arches were made of steel; he had insisted upon opening the Mississippi mouth and obtained "by his surpassing methods" an appropriation double the amount the work would cost, and even at that ruinous expense the jetties were a wretched failure. "The business of wrecking," the Admiral concluded, "seems to embody the genius of Captain Eads . . . and future Americans will hold him as the greatest wrecker known in our whole history."[9]

James Eads had read the pamphlet curiously and dismissed most of it as "offensive personalities and billingsgate," but the loyal Corthell could not let it go at that. He wrote a reply, taunting Ammen with having been much disturbed over a mere reference to the unusual ability of Captain Eads. This touched Ammen off to another pamphlet in which he protested: "The false presentation of Captain Eads no more disturbed me than would the reading of a description of 'the woolly horse' put up by the bill poster of the great showman. In case that a belief in the woolly horse by a large number of well meaning people affected the national interests, destructively to a degree, I would state my reasons and produce my testimony that it was not a woolly horse at all, but quite another animal from the one presented in terms to excite our wonder and imagination."[10]

Getting around to his bitterness over Eads having attracted General Grant away from the Nicaragua project by his enthusiastic talk of the ship-railway, Ammen said in all sincerity: "It is a melancholy fact that the skillful fowler knew how to spread his nets. . . . Going to Mexico with General Grant . . . he was able to be a principal factor in diverting him from his great purpose and imposing untold sorrows instead of allowing his closing years to be his grandest. Not alone is this deplored because General Grant's last years were made inexpressibly sad; with a European power, especially Germany, in possession of that route, a canal will be speedily constructed, the lake will then become the most important naval station on the globe, Central America will be Germanized, guns will bristle everywhere, Cuba will be purchased, and what will we do about it? Year in and year out there has stood Captain Eads and his indomitable associates to prevent an act to incorporate the Nicaragua Canal Company."[11]

James Eads pondered this extraordinary outburst. None of it made any difference except the misleading statements about his fancied conflict with General Grant. To clear these up, and to show how their friendship had weathered all minor differences, Eads published his correspondence with Grant in a New York newspaper,[12] then bent his every effort toward overcoming the scant remaining suspicion of his ship-railway plan.

Rarely had Congress had such a volume of troublesome legislation, and never had lawmakers been more fearful of executive veto. President Cleveland, now one year in office, exercised the veto power "beyond all precedent."[13] James Eads, however, felt confident that the President would sign the ship-railway bill when it passed—Cleveland had been most friendly to him in his blunt, undemonstrative manner, and he wanted an isthmus crossing. But some of the Tehuantepec ship-road enthusiasts did not trust Cleveland so entirely. An English journal complained that "the President smites with one hand and strokes with the other."[14]

The session spun on into early summer. Eunice spent much time in Washington, bringing with her her blond, buoyant daughter, Adelaide, whose marriage to General Hazard appears to have ended in separation. June brought pretty twenty-two-year-old Frances Folsom to Washington where she was married to her earlier guardian, the President. She became at once "an enthusiastic admirer of Captain Eads."[15] He escorted her on her first tour of the Capitol, and took her several times to see the ship-railway model, which was drawing many visitors to a basement room of the House of Representatives.[16] Occasionally in the late afternoon, at the end of a busy day, Eads could be seen driving, with Eunice, Adelaide and the vivacious First Lady, out Fourteenth Street and Woodley's Lane to Tennallytown Road where Mr. Cleveland had bought a decrepit house which he was going to remodel.[17] It was a fashionable drive, one saw everybody along the way, and there was much nodding and hat tipping from carriage to carriage.

James Eads was resolutely without fear for his greatest venture as he steered his bill through the intricacies of committee to the floor of Congress, and even when he saw it "passed over" there day after day. To him the ship-railway was such a logical, simple answer to the vital isthmus question that it was bound to win past the scant opposition. But when the long, hot summer had ended, the many friends of his bill had not been able to push it through, so effective had been the tactics of its few opponents.

It was tantalizing to have success, so nearly in his grasp, elude him. For seven years the Mississippi had waited for Congress to sanction a short route for its trade to the American West Coast and the Orient—how much longer would it have to wait? Even the frail hope of a distant crossing at Panama was stifled, for it was evident to all but the most credulous that the de Lesseps tide-level canal was a phantasm that would vanish. Laying his lines determinedly for another session of Congress, James Eads resolved anew that he would find a way to gain his country's approval of the ship-road that would be such a boon to it.

He was strained, but showed no despair as he divided his time between Washington, New York and Port Eads. However heavy lay his disappointment, he bore the same sturdy, graceful mien and was as ready as ever with a good story or a favorite poem. In and out of his other activities he had, all along, woven a strand of watchfulness over the jetties, brushing aside missiles of criticism that still showered upon him from some of the Army engineers. The government inspectors, obliged to please their irascible chief, General Humphreys, had from the first searched diligently for any sign of shoaling in the jetty channel, and even in the pass. Once when, by basing a survey upon a different level from that designated in the Jetty Act, a zealous inspector found four inches of depth lacking in one part of the pass, a roar of condemnation from Humphreys's office blasted away a maintenance payment that Eads was about to receive. The Secretary of War had ordered the survey made over, and again, by "an oversight," the wrong level was used. When a forced correction was made, the shoal proved to be a surplus depth. There were other times when a few inches of depth actually were lacking in the pass, and although Eads had no responsibility except within the jetties, the mild southern air of Washington was rent by accusations.[18]

All this froth and bickering made little difference to James Eads as long as the world's great ships sailed in between the willow walls to the very heart of the throbbing valley commerce. From John Ubsdell's cottage at the head of the oleander- and rose-bordered plank walk, where he stayed when in Port Eads, he could hardly glance up without seeing a ship passing in or out of the channel. A deep channel for the whole Mississippi would come, too, he was convinced. It must. And a short way to the western oceans.

In St. Louis there were always men with river interests waiting to see James Eads when he stopped there, and there were former neighbors and old friends. He came with reassurances about the future of the Mississippi, with enthusiasm, with accounts of his travels and experiences, with stories and gifts. To his grandchildren he was a chum and playmate. They adored him and his pranks, yet they stood somewhat in awe of him, they had heard much flattering comment about him and looked upon him as a great and mysterious man. To four-year-old James Eads Switzer, Mattie's little son, one of the mysteries was the skullcap his grandfather wore. One day he dared to speak about it. "Why do you always wear that cap? How would you look without it? I've never seen it off of you."

"You haven't? Then look now." Eads dodged behind a door and held the cap around the corner on his hand. His small namesake stared at it in perplexity. It vanished and came out on the head of its owner.[19] Surely, the boy must have thought, his grandfather was a mysterious man, and he kept on wondering what was hidden under that cap.

Many a surprising thing had lurked under it at one time and another, grotesque ironclads, the Bridge, willow walls that ordered the mighty Mississippi and swayed the commerce of half the earth, and ships that climbed aboard trains to ride over plains and through jungles. All these had appeared fantastic at first, all but the overland trains were commonplace enough now. And there were less tangible mysteries under the small cap, strengths and tensions, faiths and fancies, a farseeing shrewdness, a peculiar oneness with all men, an abiding consciousness of Supreme law. But the dominant thing that looked from beneath it through the direct gray eyes was fearlessness. Perhaps a major part of James Eads's genius was his unflinching courage to stride past all precedent into the untried.

Late autumn found James Eads back in New York, submerged in work. The days dragged on him, yet they were not long enough for all that he had to do. With frequent conferences, and innumerable letters written on his ship-railway pictured stationery,[20] he was preparing to make a daring sacrifice so that the Tehuantepec crossing might be realized: *He would assume the entire burden of financing the seventy-five million dollar venture,* asking nothing of the government but a sanctioning charter.[21] It would be a colossal undertaking, and he was not well, but the ship-railway must be built. The inland rivers, and even the seacoasts, were clamoring for it, while timid legislators condemned it although experts in Europe and America insisted that it was entirely practicable. In December, just as he had everything in shape for a descent upon Congress, Eads took a heavy cold. It was annoying, he had no time to be ill now, he would go on to Washington. But the doctors were grave and Eunice so alarmed that he let them persuade him to take a short rest at Lakewood, New Jersey.[22] Eunice and Adelaide nursed him devotedly, the beauty and quiet of the resort were a balm to him, he walked in the pine woods, enjoying the sharp, clean air. But he was restless. He could not stay. Against all advice and pleading, he persisted in going to Washington as the year 1887 opened.

He felt better when he got there. It was where he wanted to be. He was eager to put forward his new proposal. This would be the last stand he would ever have to make for his ship-railway, for certainly no body of lawmakers would refuse his offer to give so much in return for a simple endorsement of his plan. And this time there could be no possible grounds for a hostile attack— he had seen to that, combing his proposition over and over for months. Confidently he paved the way to have his bill introduced in the Senate.

The weather was raw and mean in Washington, the wind chilled him through, but the favorable comment upon his offer from every side was like a draught of wine to him. He basked in his certainty of success. At last the Mississippi would have its outlet to the Pacific, the eastern and western coasts would be drawn ten thousand sea miles closer together. Then, suddenly, all friendly opinion was drowned in the outcries of his opponents: Captain Eads was making a sly attempt against his country! He was slipping up on the government with an insidious scheme to throw the United States open to the mercy of the world! Oh, no, he was not asking for government financial aid, of course not—because he meant to enlist foreign capital and allow the control of the American isthmus to be grasped by European powers![23]

For a moment it seemed that James Eads might reel under this unexpected onslaught, which was all the more startling to him because he had gone to such lengths to guard the interests of his country. He came out with a sharp denial, showing that he had been "urged by a member of the British Parliament" to submit his earlier proposition to that government, and that he had refused to do so "because of the desire of the company to have only our country and Mexico carry on the work."[24]

The ship-railway could have been almost ready for use now, wiping out the barrier that had afflicted shipping for centuries, except for the time wasted upon groundless suspicion and fear, he said. No crossing was being provided at Panama, or likely would be for thirty years. Meanwhile, there sat the whole world baffled by a narrow strip of land! Where was the ingenuity that the Creator had lent men to fashion the fullest possible life from the earth's bounty? Why did laymen insist upon deciding engineering problems in a way exactly counter to the findings of engineers themselves?

With a surge of energy, James Eads turned upon the opposition. Of all the struggles he had ever made in behalf of commerce and the Mississippi, this was the most intensive. He had offered the country his all, he had asked hardly more than an acknowledgment—how could it be denied him? He was everywhere, parrying blows, meeting attacks like a skillful fencer. His shoulders were erect, his arguments penetrating, even though exhaustion pulled at him. "Full of weariness and pain" he put forth his waning strength to secure for his river and his country a short route to the Pacific, "a work to him so great and vast in its conceptions and its possible results that he often appeared to be lifted to the spiritual plane, moved by a heaven-born purpose."[25] He had never had the least doubt, after his first calculations, that the ship-railway would be feasible and serve shipping admirably, keeping pace with it as it grew. "Believe me," he had long ago written General Grant, "I would be unwilling to stake my reputation upon a doubtful experiment."[26]

Inflexibly he drove his bill forward, ferreting out every advantage, defending each foot of progress, and was rewarded by a steadily returning sentiment in favor of it. By the middle of January the House stood almost solidly behind it and a wide majority in the Senate had informally declared for it, although it had not come to a vote in either house. There seemed no doubt now that the measure would go through to victory with a flattering margin. James Eads returned to New York for a short breathing spell. He was a little tired, he admitted, but a few days of rest would set him straight. Eunice wondered if it would, for there was some new factor in his fatigue. She called a doctor, and another. The physicians recognized, beneath the obstinate will that kept this man of steel so upright and forceful, a dangerous exhaustion. They ordered him away for a rest. Rest! Now? When his ship-railway prospects were at a crisis? But rest was imperative, he was warned. Rest—how many times he had had to quit at the very height of a battle! He consented to go, it need not be so far that he could not rush back to defend the ship-railway bill if it needed him. He chose Nassau, in the Bahama Islands, and sailed on February 3, 1887, Eunice and Adelaide with him.[27]

The ocean trip relaxed him—there was nothing like an expanse of water to ease away tensions, it was always giving, yielding, seeking a level of vibrant peace, content to spend an eternity at it. The creak of the ship, the mild roll, the inconsequential talk that rose around him, were soothing. Presently he was debarking. Everything was attended to for him, he could lean back in a carriage and listen to the patter of hoofs until they drew up before a hotel.

Stretched in a chair on his balcony he could look away to the misty horizon. Yonder somewhere the brown waters of the Mississippi reached with their commerce toward the shores of Mexico, to Minatitlán where the ship-railway would begin. He had often said that he could see, with inward vision, ships borne overland on his railway, so real to him was its ultimate existence. And he would murmur or declaim:

> "Lo, ships, from seas by nature barred,
> Mount along wayes by man prepared,
> Along far-stretching vales whose streams
> Seek other seas, their canvas gleams."[28]

He could see them now, stately vessels, being raised from the sea by the lift he had invented—he had patented four inventions for the ship-railway, the last one being the fifty-first American patent he had taken out. This was a positive hydraulic lift, powerful and efficient, to take the place of the caisson lift that he had earlier favored for its economy. Lately he had contrived improvements for his lift and was awaiting the detailed drawings of the changes that he might examine them.[29]

His strength came back stingily, the days ticked slowly by. He grew restless. Something might go wrong in Washington, he told Eunice, and he had better return. If the opponents of the ship-railway should make some new attack, he must be on hand to defend it. No telling what further stories they had concocted. Eunice reasoned with him, coaxed him. He put this gently aside. He was going home. Reservations were made on a Ward Line steamer for February 24.[30] It was hard for Eads to wait for the sailing date, so much was at stake, but the drawings of the new lift came and he began, with his hungry energy, to check them, forgetting that he must rest. A fresh cold gripped him, he was put to bed, but the lift plans went with him. Propped against pillows, he pored over them, approving them here, changing them there. It was absorbing work, but in the midst of it the pencil dropped from his fingers, his tired body sank back, he gasped for breath.

A dispatch telling him that his bill had passed in the Senate gave him a new spark of life.[31] There was no doubt now that the bill would become law, for the house had been almost solidly in favor of it. The ship-railway would be realized, the seas of the world linked together. He had only to get well enough to go with all his might at building it. Commerce would not have to wait upon the far-off completion of a Panama crossing.

James Eads was never to know that his ship-railway bill would not have even a chance to be voted upon in the House of Representatives, owing to "The arbitrary and injurious action of Speaker Carlisle in denying the few minutes necessary for its consideration."[32] Fatigue that had dogged the steps of the overworked man so long was crowding him aside. Time, he pleaded, just a little more time. He had often said, "I shall not die until I accomplish this work and see with my own eyes great ships pass from ocean to ocean over the land."[33] A while longer, a little while. The mist was moving in from the horizon, it filled the room with a strange twilight. "I cannot die," the worn man said firmly, "I have not finished my work."[34]

But for the valiant champion of the Mississippi time was merging into the boundless point of eternity, his stubborn will was slipping into the sea of universal will. On March 8, 1887, the great engineer closed his eyes.

The ship-railway, which had pulsed with his every heartbeat, stood apart to live on after him.

CHAPTER XIX

THE RIVER ROLLS ON

A SMALL steamer, the *Lizzie Henderson,* brought the word to Key West, where it could be flashed over the nation's telegraph wires: James Eads had died, and his work not yet done. The news was a profound shock to all but his most intimate circle. It had not been generally known how spent he was. Dismay over the loss of the man who, it was believed, would surely furnish a shipway across Tehuantepec was nearly as deep in Boston and Brooklyn as along the inland rivers. It was widely felt that he had been sacrificed to a bitter opposition by the inertia of a friendly public who expected him to provide, somehow, the only crossing of the isthmus for which there seemed the slightest hope. The Augusta (Georgia) *Chronicle* put it in these terse sentences: "It is melancholy to think that mousing owls should be permitted to strike down such an eagle. . . . The greatest engineer in the world is dead."[1]

At Port Eads the pilot station, tugboats and white cottages flew flags at half-mast. In New Orleans keen regret reigned. The port had staked exuberant hopes on the ship-railway, firmly convinced that its projector would finally tower above his opponents and build it—now his dynamic energy was stilled. But it was of their gratitude to James Eads, rather than of their stricken hopes, that people talked on their galleries in the evening, and of how he had won his way past their coldness and suspicion into their hearts.

"New Orleans can never forget . . . that it ridiculed the idea that any man could bridle the current of the Mississippi, and generally opposed the jetty proposition," the New Orleans *Times-Democrat* gave words to a prevailing regret. "Captain Eads, by opening the river . . . has already added $1,800,000,000 to the wealth of the farmers and manufacturers of this vast region which he opened to the largest steamers of the world."[2]

Businessmen held meetings to plan a demonstration of respect, their flattering port statistics were repeated, anecdotes told, and memorials of Eads in bronze or marble suggested.[3] New Orleans had, indeed, lost a friend.

In less demonstrative St. Louis there was an air of waiting silence. The body of James Eads was brought home on a fitful spring day, March 17. A few persons watched bleakly as the small cortege of family carriages wound southward toward Lafayette Park to draw up at the home of Eads's older daughter, Eliza How.[4] And there Mattie and Josephine and their husbands gathered. Eunice was stricken, but self-possessed. She was used to partings, to being alone, but there had always been expectancy. She groped for something to take its place. He was off on another trip, without baggage, but too much had clung to him and gone with him—his strangely transcendental yet practical philosophy, his air of confidence and well-being which enveloped those about him, his mingle of protectiveness and childish dependence. His daughters and hers were grieving over him with equal abandon, he had made no difference between them.

News that the remains of Captain Eads had been brought back to St. Louis passed along the streets and down the levee. Men of every rank made their way to the How home to offer their respects before the bier in the east parlor, or to stand silent outside the house. Some drove or trudged past the mansion on Compton Hill. The garden still lay under a winter hush. There on the lawn Eads had liked to sit, looking away to where the city lights twinkled against the night sky. With a court around him he would spin his yarns and laugh delightedly over them, explain why the long Bridge arches did not crumble, or how flimsy willow walls could discipline the unruly river. It seemed incredible that he was gone—he had visited St. Louis only three months ago and spoken, with eyes alight, of the growing Mississippi commerce, confident that it would have a short route to the West Coast and the Orient.

On the next afternoon, while tolling bells punctuated the journey to Christ's Episcopal Cathedral, men who had been too shy or too humble to edge into the house on Lafayette Avenue gathered in the streets—one-time workers in the Carondelet shipyard or on the Bridge, steamboat crewmen or wharf hands, holding their caps as the long line of carriages rumbled slowly past. After a simple service, the funeral procession wound across town to Bellefontaine Cemetery, spread on the summit of the double limestone terrace that rose from the river. The St. Louis Engineers' Club followed the casket in a body.[5]

At a public memorial meeting held on the following day at the Merchants' Exchange, Mayor David R. Francis, Henry Flad and many others offered spontaneous tributes of respect. It was not primarily of James Eads's great engineering feats that they spoke, but of the little intimate things about him, his big heart, his manifold deeds of kindness. "How many mouths he fed and bodies he clothed, how many . . . words of encouragement he spoke to the struggling, God only knows. . . . A thousand times have I heard him urge excuses for those who offended him," one speaker said. "He had the courage of a lion and the gentleness of a woman," another commented.[6] The listeners recalled to each other that Captain Eads had made the last years of his early employer, Barrett Williams, comfortable and contented . . . he had spent part of a belated government payment on the gunboat *Benton* in war relief . . . he had given lavishly to ease the plight of southern refugees who had swarmed here.

Small ports and landings along the Mississippi wondered who would now carry the banner for the river, plan for it, fight for it. All chance of an opening to the Pacific for its commerce seemed shattered, now that James Eads was gone. Even the distant crossing at Panama was a dimming mirage.

There were a few determined friends of the Eads Tehuantepec plan, however, who meant to carry it through, statesmen, engineers and businessmen who saw in the ship-railway the only hope of conquering the isthmus barrier within their time. A group of these, led by ex-Secretary of the Treasury William Windom, William Thaw, Jr., and John H. Rice, gathered on a June day, in 1888, at 32 Nasasu Street, in New York City, to organize a new Atlantic and Pacific Ship-Railway Company which they had incorporated.[7]

"The prediction made by Mr. Eads in 1880 in reference to the almost insurmountable difficulties at Panama have been realized," they opened their discussion. Another way between oceans would have to be provided. The history of the ship-railway venture was sketched, old and new letters from eminent engineers endorsing the Eads plan were read, a contract with James Andrews, Eads's associate in building the Bridge and the South Pass jetties, to head the construction of the ship-road was signed, and the Mexican concession was bought from the New Jersey company that Eads had formed to hold it. Here, departing from the procedure of James Eads, who, although he fanatically wanted the Tehuantepec ship-road to materialize, had refused to appeal to any country but his own and neighboring Mexico for aid, the directors detailed William Windom to sail for Europe and arrange finances wherever he could.[8]

It happened that during this year, 1888, the prospect of a canal at Panama grew brighter for a while. There actually had been obtained a fifteen-mile waterway, beginning at Colón on the Atlantic side, deep enough to float a thousand-ton vessel. Investors held off from any crossing that would be in competition with it. But the avalanche of ruin that had begun to slide upon the French undertaking could not be stayed. The venture was half buried under incompetence and waste. Its funds were exhausted, and a lottery loan, proposed by de Lesseps, failed to replenish them.[9] Again American public interest swayed to the Tehuantepec ship-railway, and again it was diverted, this time by the proposed Nicaragua canal.

Now that James Eads no longer "stood there" to block them, the heads of the Nicaragua company managed, early in 1889, to get a charter bill through Congress. President Cleveland signed it just as the bankrupt Panama company were laying off their workmen, leaving much of their machinery idle. The Nicaragua canal backers hastened to buy a large amount of the equipment from the impoverished Panama works and brought it to the Nicaraguan isthmus. In the following two years about four million dollars was spent on a canal there, then the project languished. But its early rapid progress had given the Tehuantepec ship-railway another backset.

Disillusion crept like a paralysis over the two American attempts. The French effort was gasping out. Paris seethed with indignation over the Panama fiasco. The French government opened an inquiry. The report, made in November, 1892, was a nearly incredible story of inspired enthusiasm, public trust, reckless spending, toil, fever and corruption. There were sensational scenes in the French Parliament, accusations of bribery of legislators by the Panama company, challenges to duels. The venerable Viscount de Lesseps was convicted of fraud and confined to his country home, where he died in 1894. A new Panama company, risen from the ashes of the old one, went on chopping futilely at the strip of land.[10]

The American isthmus still plagued the maritime world, steamers and sailing vessels made the long voyage around Cape Horn from ocean to ocean while men dreamed vaguely of a shipway across the barrier, and the Mississippi Valley pondered wistfully what might have been if James Eads had not been taken so untimely. A veteran steamboat man, Captain Emerson Gould, who had opposed the St. Louis Bridge with all his might, wrote: "If Captain Eads had lived there is little doubt but what he would have built a ship-railway across the isthmus. . . . There has never been but one James Eads in America."[11]

Now and then advocates of the Tehuantepec ship-railway, with Elmer Corthell in the lead, tried vainly to finance or even charter their project. In December of 1895 a strong written argument by Corthell was put before Congressional committees,[12] but nothing came of it. Nine years later, in 1904, after a long and acrimonious debate over the advantages of the respective routes, the United States bought the Panama holdings of the discouraged French company and began work on a lock, not tide-level, canal.[13]

James Eads had said, when Ferdinand de Lesseps began his labors at the isthmus, that it would be more than thirty years before a canal could be realized. Thirty-four years after this prediction was made, the lock canal at Panama was opened to traffic, on August 15, 1914. At last there was a shipway across the isthmus, but some Americans still thought that the failure to build the ship-railway at Tehuantepec, at least to fill in the weary time before the canal was ready, had been a grave mistake. Six years after the completion of the canal an article in an engineering journal commented:

"What might have been the history of isthmian transportation if the ship-railway had been built can only be a matter of conjecture. It may be that engineers will again consider the possible application of the ship-railway for transportation in commercial and military service."[14]

Both engineers and statesmen have again considered it. In 1939, after Secretary of War Harry Hines Woodring had requested Congress to spend a hundred and forty million dollars to construct another and larger set of locks at Panama, since estimates had set forth that the present facilities would be taxed to the utmost by 1961, Secretary of the Navy Charles Edison proposed a ship-railway at Nicaragua.[15] The fear with which some legislators had viewed James Eads's proposal to lift loaded vessels from shore basins forty-six feet to a railroad track had long ago been proved unfounded, for such a feat had become commonplace—the improved hydraulic docks at Malta, Bombay and Singapore could each raise fifty thousand tons, the docks at Southampton could raise sixty thousand.[16] And the hauling of a loaded ship overland no longer appeared to be a task that would outgrow motive power, for railroad engines had steadily gained in size and strength, as Eads had insisted they would, some of them now of giant power. Congress, however, finally decided that an added set of locks at Panama should be constructed.[17]

But the idea of a ship-railway across the isthmus has never been vanquished. In August of 1946, Señor Modesto C. Rolland, engineer, architect, and official in the Mexican government, a man of spectacular achievements as a builder, endorsed a plan for a ship-railway over the Tehuantepec route that James Eads had adopted, from Salina Cruz on the Pacific to the Coatzacoalcos River emptying into the Atlantic, using powerful Diesel-motored locomotives to draw the ship-car overland. His plan includes having Mexico hold fifty-one per cent of the capital stock, with possibly the United States and Britain subscribing the rest.[18]

Sixty years after James Eads's death, his works are living and serving. Engineers still marvel at the St. Louis Bridge. "Even in an age which has seen the erection of great new bridges . . . it continues to rank among the outstanding bridges of the world." The late President Samuel W. Stratton of the Massachusetts Institute of Technology said he considered it the most beautifully lined bridge he had ever seen."[19] Over the vehicular roadway on its top, lately widened and given an ideal surface by the Terminal Association of St. Louis, owners of the Bridge, a million passengers by bus alone, and about three hundred thousand private vehicles, cross the river each month. The trip over the Bridge highway is an unforgettable scenic excursion.[20]

Rock-bolstered brush mattress walls still maintain a deep seaway at the mouth of the Mississippi. After the South Pass jetties were completed, delays of water traffic there were unknown. But ships were outgrowing the little pass by the end of two decades, as Eads had prophesied they would, and in 1902 Congress broke across a debate over an American canal at Panama to authorize the creation of a deeper, wider channel at large Southwest Pass. Seven years later this channel was completed. None of the hostility and suspicion that had beset James Eads were on hand during this construction, no chorus about increased bar advance, for the South Pass jetties had cut the pushing out of the bar from a hundred and eleven feet a year down to forty-four feet during the sway of Eads over them—a later change made in the old jetties raised the bar advance to a hundred and twenty-one feet annually,[21] but this could be avoided in the new ones. Louisiana considered the Southwest Pass jetties Eads's too, for the concrete-capped willow mattress walls he had designed were used, only slight changes being made in his plan.

New Orleans felt that James Eads was still protecting its port, and in October, 1929, staged a two-day Golden Jubilee in honor of him and his work. The whole state entered into it, the Governor declared a Navy Day, the U.S.S. flagship *Rochester* was brought up from the Canal Zone to open the ceremonies. Aboard it a party went to Port Eads, followed by others on a boat named, by quirk of fate perhaps, the *General Humphreys*. At Port Eads a bronze memorial tablet was unveiled after school children had laid a wreath at its base, and twenty-one guns were fired. The celebration moved back to New Orleans, where thousands stood in a pouring rain to watch a river procession of flag-draped vessels, hydroplane and motorboat races, street parades, and to hear speeches by the Governor and by men who had known Captain Eads. Another tablet was unveiled.[22]

When the Panama Canal provided the Mississippi with at least a distant trade outlet and made it a world river, the condition of its navigation channel became more important than ever. Off and on, limited or ambitious attempts have been made to improve the channel, with slight or substantial success. The jettying of the shoaled stretches, fought for so resolutely by James Eads, has been alternately abandoned and readopted. The Army engineers in charge generally echo his insistent declaration: "Levees are not enough, and patchwork is wasteful." Their plans, like his, are largely based upon forcing the river to do much of the work.[23] A navigation channel with a least depth of nine feet at low water now exists down the Mississippi's whole navigable length.

The great value of the Mississippi River system in time of war, which James Eads stressed so tenaciously, was to be dramatically shown by the enormous transportation task that its five thousand barges and one thousand towboats performed in World War II, the largest towboats being able to push as much freight at once as three hundred railroad freight cars could carry.[24] And where landsmen had been astounded at the grotesque ironclads set afloat from the Carondelet shipyard, folk eighty years later were to grow accustomed to seeing the most outlandish vessels, LSTs, submarines and other monsters of war, built on the inland streams, headed southward on their way to the Mediterranean or the South Pacific. But the river's greatest transportation service has always been its peacetime traffic, with its lowering of all types of haulage rates.

James Eads had believed fervently that cheap transportation was the prime need of this country of long distances. Early in his life, addressing the students of a St. Louis trade school, he had said: "The man who can by his inventive genius cheapen light or heat or the supply of water to the masses is greater than a general. . . . He who by his genius can cheapen bread or clothing to the poor, or any of the common necessities of life, as sugar, salt, soap or fuel, deserves the thanks of the nation, and to have his name inscribed in the temple of fame."[25]

As if he had spoken the words about himself, a place in the American Hall of Fame, in New York City, was accorded him, the first engineer to be thus honored, in 1920. And four years later a bust of him was unveiled there, his grandson, James Eads Switzer, taking part in the ceremony.[26]

High on the crest of the terraced bluff face of St. Louis, in beautiful Bellefontaine Cemetery, not far from where rest Henry Shreve, Thomas Benton, and pathfinder William Clark, James Eads sleeps while his achievements live on. The powerful armored warships of today reflect distantly the light of his findings. No large bridge of deep foundations and steel superstructure is stretched over water but rests, remotely at least, upon his pioneering. Rivers in many a country pour untrammeled to their outlet because of the South Pass jetties. Lapping against the foot of the bluff on which he lies, the broad stream to which Eads devoted most of his life, and for which he died, pushes on, lunging past prairie and hill, plantation and city, on its way to the seven seas.

NOTES

CHAPTER I

[1] Charles Brandon Boynton, *A History of the United States Navy During the Rebellion*, vol. I: biographical sketch of James B. Eads; Charles E. Snyder, "The Eads of Argyle," *Iowa Journal of History and Politics*, vol. 42, no. 1.

[2] Louis How, *James B. Eads*.

[3] Snyder, "The Eads of Argyle," *Iowa Journal of History and Politics*, vol. 42, no. 1.

[4] Archibald Henderson, "Eads, Master Engineer," *Universal Engineer*, vol. 55, no. 1 (1932).

(Remark: Dr. Henderson states that deans of engineering schools had selected as the five greatest engineers of all time, Leonardo da Vinci, James Watt, Ferdinand de Lesseps, James B. Eads, and Thomas A. Edison.)

[5] Estill McHenry, *Addresses and Papers of James B. Eads*; "James B. Eads," *Central Magazine*, July, 1874.

[6] "Sketch of James B. Eads," *Popular Science Monthly*, vol. 28; Floyd C. Shoemaker, *Missouri's Hall of Fame:* James B. Eads.

Locations and descriptions of all St. Louis places mentioned were furnished by Dr. William G. Swekosky, St. Louis.

[7] Same as note 6.

[8] J. N. Taylor and M. O. Crooks, *Sketch Book of St. Louis*; Robert Baird, *A View of the Valley of the Mississippi;* James Stuart, *Three Years in North America;* John Thomas Scharf, *A History of St. Louis;* Thomas Edwin Spencer, *The Story of Old St. Louis*.

[9] James Eads Switzer, "The River," a poem.

[10] Richard Edwards and M. Hopewell, *The Great West and Her Commercial Metropolis, St. Louis*.

[11] "James B. Eads," *Central Magazine*, July, 1874.

[12] *Ibid.*

[13] *Ibid.;* New York *Daily Tribune*, Sept. 26, 1880; Snyder, "The Eads of Argyle," *Iowa Journal of History and Politics*, vol. 42, no. 1.

[14] Records of Bellefontaine Cemetery, St. Louis, furnished by Francis C. Burgess, Supt.

[15] August P. Richter, "Le Claire is Interesting," Des Moines *Register and Leader*, Aug. 2 and 9, 1924; Des Moines *Register and Leader*, Apr. 18, 1909, loaned by Mrs. Frank Gordon, Le Claire, Iowa; Snyder, "The Eads of Argyle," *Iowa Journal of History and Politics*, vol. 42, no. 1.

CHAPTER II

[1] Edwards and Hopewell, *The Great West;* Helen Davault Williams, "Social Life in St. Louis, 1840-60," *Missouri Historical Review,* vol. XXXI, no. 1; Dorothy B. Dorsey, "The Panic and Depression of 1837-43," *Missouri Historical Review,* vol. XXX, no. 2.

[2] Fred Erving Dayton, *Steamboat Days;* Frank Haigh Dixon, "A Traffic History of the Mississippi System," *National Waterways Commission,* Doc. 11; Elisabeth McClellan, *Historic Dress in America.*

[3] Edwards and Hopewell, *The Great West;* "Do You Know That," *Missouri Historical Review,* vol. XXXII, no. 3; George Washington Smith, *A History of Illinois.*

[4] "James B. Eads," *Central Magazine,* July, 1874; Helen D. Williams, "Social Life in St. Louis," *Missouri Historical Review,* vol. XXXI, no. 1.

[5] Emerson Gould, *Fifty Years on the Mississippi;* letters of James and Martha Eads loaned by James Eads Switzer, grandson of James B. Eads; descriptions from photographs furnished by Mr. and Mrs. James Eads Switzer, and Missouri Historical Society, St. Louis.

[6] Gould, *Fifty Years on the Mississippi;* George B. Merrick, *Old Times on the Upper Mississippi.*

[7] J. D. Barnes, "Old Times on the Mississippi, with Personal Recollections of Early Days," Port Byron (Ill.) *Globe,* in the 1890's, date missing, from collection of Mrs. Frank Gordon, Le Claire; Snyder, "The Eads of Argyle," *Iowa Journal of History and Politics,* vol. 42, no. 1; Bellefontaine records, by Francis C. Burgess, Supt.; Maria L. Follett, "Our Pioneer Women" (Address before Settlers' Association, Oct. 4, 1897.)

[8] Gould, *Fifty Years on the Mississippi.*

[9] *Ibid.;* Taylor and Crooks, *Sketch Book of St. Louis;* "James B. Eads," *Scientific American,* vol. LVI, no. 17.

[10] "Sketch of James B. Eads," *Popular Science Monthly,* vol. 28.

[11] Davenport *Gazette,* Nov. 28, 1850; James B. Eads, *Report on the Physics and Hydraulics of the Mississippi River.*

[12] James B. Eads, *Report on the Physics and Hydraulics of the Mississippi River.*

[13] Snyder, "The Eads of Argyle," *Iowa Journal of History and Politics,* vol. 42, no. 1; Maria L. Follett, "Our Pioneer Women."

[14] Letters of James and Martha Eads, loaned by James Eads Switzer.

[15] *Ibid.*

[16] *Ibid.*

[17] *Ibid.;* records furnished by Dr. William G. Swekosky.

[18] Letters of James and Martha Eads, loaned by James Eads Switzer.

[19] *Ibid.*

NOTES 309

[20] Scharf, *A History of St. Louis.*
[21] Letters of James and Martha Eads, loaned by James Eads Switzer.
[22] Church records furnished by Rt. Rev. Msgr. John P. Cody, Chancellor of the Archdiocese of St. Louis; data, Dr. William G. Swekosky.

CHAPTER III

[1] Data by Dr. William G. Swekosky.
[2] (Given in St. Louis Directory as at Broadway and Jefferson, which would now be Broadway and Clinton, in North St. Louis.) Records by Dr. William G. Swekosky.
[3] Thomas Knox, *Robert Fulton.*
[4] Logan U. Reavis, *St. Louis, the Future Great City of the World;* Taylor and Crooks, *Sketch Book of St. Louis.*
[5] Rolla Wells, *Episodes of My Life;* Helen D. Williams, "Social Life in St. Louis," *Missouri Historical Review,* vol. XXXI, no. 1.
[6] *Ibid.;* Scharf, *A History of St. Louis.*
[7] Henry Inman, *The Old Santa Fe Trail;* Justin Henry Smith, *The War with Mexico.*
[8] Letters of James and Martha Eads, loaned by James Eads Switzer; Bellefontaine Cemetery records.
[9] Same as note 8.
[10] Louis How, *James B. Eads;* Letters of James and Martha Eads.
[11] Taylor and Crooks, *Sketch Book of St. Louis.*
[12] Bellefontaine Cemetery records.
[13] From poem, "Lines to an Absent Husband," by Martha Eads, in the Davenport *Gazette,* Aug. 17, 1848, Davenport Public Library; letters of James and Martha Eads, loaned by James Eads Switzer.
[14] Des Moines *Register and Leader,* Apr. 8 and Aug. 9, 1924.
[15] Scharf, *History of St. Louis;* Reavis, *St. Louis, the Future Great City;* Spencer, *Story of Old St. Louis.*
[16] Bellefontaine Cemetery records.
[17] Elmer L. Corthell, *James B. Eads* (pamphlet).
[18] Letters of James and Martha Eads, loaned by James Eads Switzer.
[19] Taylor and Crooks, *Sketch Book of St. Louis;* Gould, *Fifty Years on the Mississippi.*
[20] Letters of James and Martha Eads, loaned by James Eads Switzer.
[21] Davenport *Gazette,* March 27, 1851, Davenport Library.
[22] Helen D. Williams, "Social Life in St. Louis," *Missouri Historical Review,* vol. XXXI, no. 1.
[23] Letters of James and Martha Eads, loaned by James Eads Switzer.
[24] James B. Eads, *Physics and Hydraulics of the Mississippi River.*

[25] Letter of James B. Eads to his daughters, read to me by his grandson, James Eads Switzer.
[26] *Ibid.*
[27] Spencer, *Story of Old St. Louis.*
[28] Data by A. D. Stevens, grandson of Sue Dillon Stevens.

CHAPTER IV

[1] Letters of James and Martha Eads, loaned by James Eads Switzer.
[2] Data by Wallace McHenry, of St. Louis, grandson of Eunice Hagerman Eads, second wife of James B. Eads.
[3] Copy of church record furnished by Rt. Rev. Msgr. John P. Cody, Chancellor of the Archdiocese of St. Louis.
[4] Data by Wallace McHenry, grandson of Eunice Eads.
[5] (Fifth and Myrtle Streets of that day now Broadway and Clark Avenue, from records, by Dr. William G. Swekosky.)
[6] Taylor and Crooks, *Sketch Book of St. Louis.*
[7] Louis How, *James B. Eads;* data by A. D. Stevens.
[8] Reavis, *St. Louis, the Future Great City.*
[9] Walter Barlow Stevens, *St. Louis, the Fourth City.*
[10] Taylor and Crooks, *Sketch Book of St. Louis.*
[11] *Ibid.*
[12] Joseph Mills Hanson, *Conquest of the Missouri.*
[13] *Ibid.*
[14] Calvin Milton Woodward, *A History of the St. Louis Bridge.*
[15] Taylor and Crooks, *Sketch Book of St. Louis.*
[16] U.S. Patent Office records, furnished by Chief Clerk J. A. Brearley.
[17] "James B. Eads," *Central Magazine,* July, 1874.
[18] Data on the James B. Eads mansion, compiled by Dr. William G. Swekosky.
[19] Description by Dr. Swekosky, Wallace McHenry, and A. D. Stevens.
[20] Data by Wallace McHenry.
[21] Scharf, *History of St. Louis.*
[22] Taylor and Crooks, *Sketch Book of St. Louis.*

CHAPTER V

[1] James B. Eads, "Address before a Convention for the Improvement of the Mississippi, Feb. 12, 1874," Estill McHenry, *Addresses and Papers of James B. Eads.*
[2] Emerson Gould, *Fifty Years on the Mississippi.*
[3] Sketches of Eads's friends, *Dictionary of American Biography;* Rolla Wells, *Episodes of My Life;* Scharf, *History of St. Louis.*

[4] Data by Dr. William G. Swekosky.

[5] Walter B. Stevens, "Missouri Centennial," quoting Governor Stewart of Missouri, *Missouri Historical Review*, vol. IX, nos. 3 and 4.

[6] Galusha Anderson, *Story of a Border City During the Civil War*.

[7] *Ibid*.

[8] Boynton, *A History of the U.S. Navy During the Rebellion;* James B. Eads, "Recollections of Foote and the Gunboats," in *Battles and Leaders of the Civil War*, vol. I; letter of Attorney General Bates to James B. Eads, Apr. 17, 1861, Missouri Historical Society Eads Papers, St. Louis.

[9] James B. Eads, "Recollections of Foote and the Gunboats."

[10] *Ibid*.

[11] *Ibid.;* Boynton, *History of the U.S. Navy During the Rebellion*.

[12] William E. Smith, *The Francis Preston Blair Family in Politics*.

[13] Ulysses S. Grant, *Personal Memoirs of U. S. Grant*.

[14] *Ibid.;* James B. Eads, letter to Secretary of War, May 11, 1861, *Official Records of the Union and Confederate Armies*, series II, vol. 1; Anderson, *Story of a Border City;* Thomas L. Snead, *The Fight for Missouri;* Scharf, *History of St. Louis;* Rolla Wells, *Episodes of My Life*.

[15] *Dictionary of American Biography;* Dr. William Taussig, "Personal Recollections of Gen. Grant," *Missouri Historical Society Collections*, vol. II, no. 3.

[16] James B. Eads, letter to Secretary of War, May 11, 1861, *Official Records of the Union and Confederate Armies*, series II, vol. 1.

[17] Anderson, *Story of a Border City;* Rolla Wells, *Episodes of My Life*.

[18] James B. Eads, letter to Secretary of the Navy Welles, May 14, 1861, Missouri Historical Society Eads Papers.

[19] Anderson, *Story of a Border City*.

[20] Boynton, *History of the U.S. Navy During the Rebellion*.

[21] *Ibid.;* James B. Eads, "Recollections of Foote and the Gunboats."

[22] *Ibid*.

[23] Boynton, *History of the U.S. Navy;* Louis How, *James B. Eads; Official Records of the Union and Confederate Navies*, vol. 22; Francis Trevelyan Miller, *Photographic History of the Civil War*, vol. I. (Remarks: Name of the *St. Louis* was changed to *De Kalb* under Navy control; names of the other ironclads built at St. Louis at this time were *Louisville, Pittsburg, Carondelet* and *Benton;* at Mound City, *Mound City, Cincinnati* and *Cairo*.)

[24] Letter of James B. Eads to Frank P. Blair, Oct. 29, 1861; of Barton Able to Eads, Oct. 30, 1861; of Eads to General Meigs, Oct. 29, 1861; of Eads to Attorney General Bates, Nov. 10, and of Bates to Eads, Dec. 2, 1861, Missouri Historical Society Eads Papers.

[25] James B. Eads, "Recollections of Foote and the Gunboats."

[26] Telegram from Foote to Meigs, Oct. 26, 1861, *Official Records of the Union and Confederate Navies*, series I, vol. 22.
[27] Boynton, *History of the U.S. Navy*.
[28] Letter of Attorney General Bates to Eads, Dec. 2, 1861, Missouri Historical Society Eads Papers.
[29] James B. Eads, "Recollections of Foote and the Gunboats."
[30] *Official Records of the Union and Confederate Navies*, series I, vol. 22.
[31] *Ibid.*
[32] James Mason Hoppin, *Life of Andrew Hull Foote;* printed letter of James B. Eads to General M. C. Meigs, Jan. 27, 1862, Missouri Historical Society Eads Papers.
[33] *Official Records of the Union and Confederate Navies*, series I, vol. 22.
[34] Miller, *Photographic History of the Civil War*, vol. I.
[35] Hoppin, *Life of Andrew Hull Foote.*
[36] *Ibid.*; Boynton, *History of the U.S. Navy.*
[37] Scharf, *History of St. Louis;* Anderson, *Story of a Border City.*
[38] Louis How, *James B. Eads.*
[39] Boynton, *History of the U.S. Navy.*
[40] James B. Eads, "Recollections of Foote"; Eads, "Letter to the St. Louis *Missouri Democrat*, June 4, 1867," McHenry, *Addresses and Papers;* letter of Frank P. Blair to Eads, Missouri Historical Society Eads Papers.
[41] James B. Eads, "Recollections of Foote and the Gunboats."
[42] King, J. W., U.S.N., *Report to the Navy Department on the Eads Steam Turret, Apr. 30, 1864* (pamphlet).
[43] James B. Eads, "Recollections of Foote and the Gunboats."
[44] *Ibid.*; Hoppin, *Life of Andrew Hull Foote;* Boynton, *History of the U.S. Navy;* letter of Foote to Meigs, *Official Records of the Union and Confederate Navies*, series I, vol. 22.
[45] James B. Eads, "Recollections of Foote and the Gunboats."
[46] *Ibid.*
[47] A. D. Stevens, Wallace McHenry, and Dr. William G. Swekosky, description of the Eads mansion.
[48] Louis How, *James B. Eads.*
[49] James B. Eads, "Remarks at an alumni banquet, Washington University, St. Louis, Feb., 1868," McHenry, *Addresses and Papers of James B. Eads.*
[50] Louis How, *James B. Eads;* New Orleans *Times-Democrat*, March 11, 1887.
[51] Loyall Farragut, *Life of David Glasgow Farragut;* Boynton, *History of the U.S. Navy.*
[52] Curtis B. Rollins, "Some Impressions of Frank P. Blair," *Missouri Historical Review*, vol. XXIV, no. 3; David P. Croly, *Seymour and Blair;*

William E. Smith, *The Francis Preston Blair Family in Politics; Dictionary of American Biography*: Benjamin Gratz Brown.

[53] Same as note 51.

[54] Albert Phelps, *Louisiana, A Record of Expansion.*

[55] John T. Headley, *Farragut and Our Naval Commanders;* Linus Pierpont Brockett, *Our Great Captains:* Colonel Charles Ellet; Alfred W. Ellet, "Ellet and His Rams," *Battles and Leaders of the Civil War,* vol. I.

[56] Miller, *Photographic History of the Civil War,* vol. I. (The *Cincinnati* was raised, put in service, and sunk again.)

[57] Hoppin, *Life of Andrew Hull Foote.*

CHAPTER VI

[1] Loyall Farragut, *The Life of David Glasgow Farragut.*

[2] *Ibid.*

[3] *Ibid.;* Hoppin, *Life of Andrew Hull Foote;* report by Farragut to Navy Department, *Official Records of Union and Confederate Navies,* series I, vol. 21; letter of Seth Phelps, series I, vol. 24; letter of Porter, series I, vol. 21.

[4] Letter of Eads to Assistant Secretary of the Navy Fox, Jan. 31, 1863, *Official Records of Union and Confederate Navies,* series I, vol. 24.

[5] *Official Records of Union and Confederate Navies,* series II, vol. 1.

[6] Letter from Caroline Foote to Eads, July 22, 1862; letter of Foote to Eads, July 23, 1862; letter of Baron Gerolt to Eads, Feb. 28, 1863; letter of Attorney General Bates to Eads, March 23, 1863; all in Missouri Historical Society Eads Papers.

[7] James B. Eads, "Recollections of Foote and the Gunboats."

[8] J. W. King, U.S.N., *Report on Eads Steam Turret.*

[9] (There was an Eads turret on the *Winnebago,* according to J. W. King, U.S.N., and a letter of Assistant Secretary of the Navy Fox to Porter, November, 1863, *Official Records of Union and Confederate Navies,* series I, vol. 25; there was an Eads turret on each of the *Chickasaw* and *Milwaukee* Eads said in "Recollections of Foote"; letter of Porter to Fox said that the *Kickapoo* had one turret with Eads's plan for running out the guns, *Official Records,* series II, vol. 1.)

[10] Anderson, *Story of a Border City;* Rowland Gibson Hazard, article on U.S. economy, New York *Times,* Feb. 11, 1863.

[11] Anderson, *Story of a Border City.*

[12] Louis How, *James B. Eads.*

[13] Letter of Bates to Eads, Dec. 4, 1863, Missouri Historical Society Eads Papers.

[14] Louis How, *James B. Eads;* letter of Bates to Eads, March 3, 1864, Missouri Historical Society Eads Papers.
[15] Scharf, *History of St. Louis.*
[16] Letter of Farragut to Fleet-Engineer Shock, March 6, 1864, *Official Records of Union and Confederate Navies,* series I, vol. 21.
[17] Anderson, *Story of a Border City.*
[18] U.S. Patent Office records.
[19] *Central Magazine,* July, 1874.
[20] Report of Farragut to Navy Department, May 3, 1864, *Official Records of Union and Confederate Navies,* series I, vol. 21.
[21] *Official Records of Union and Confederate Navies,* series II, vol. 1.
[22] Foxall Parker, *The Battle of Mobile Bay;* Oliver A. Batcheller, "The Battle of Mobile Bay," *Magazine of History,* vol. LV, no. 6.
[23] Boynton, *History of the U.S. Navy;* A. T. Mahan, *The Navy in the Civil War.*
[24] Scharf, *History of St. Louis.*
[25] Data by Wallace McHenry and Dr. William G. Swekosky.
[26] Scharf, *History of St. Louis.*
[27] Stevens, *St. Louis, the Fourth City;* James B. Eads, "Address of Welcome to President Johnson," McHenry, *Addresses and Papers.*
[28] Stevens, *St. Louis, the Fourth City.*
[29] Louis How, *James B. Eads;* letter of Baron Gerolt to Eads, Apr. 11, 1863 (?), and letter of Lieutenant Shock, U.S.N., to Eads, October, 1863, Missouri Historical Society Eads Papers.
[30] James B. Eads, "Address at River Improvement Convention, St. Louis, February, 1867," McHenry, *Addresses and Papers,* also pamphlet.
[31] Data by Dr. William G. Swekosky and A. D. Stevens.
[32] St. Louis *Missouri Democrat,* Dec. 4, 1867.
(Remark: The wineglass mentioned is in the Eads Collection at Missouri Historical Society, St. Louis.)
[33] Ex. Doc. No. 327, H.R., 40th Congress, 2nd session. Also published as book.
[34] *Ibid.*

CHAPTER VII

[1] Hanson, *Conquest of the Missouri.*
[2] Woodward, *History of the St. Louis Bridge.*
[3] Reavis, *St. Louis, the Future Great City;* Stevens, *St. Louis, the Fourth City.*
[4] Denton Jaques Snyder, *The St. Louis Movement;* Clem Forbes, "The St. Louis School of Thought," *Missouri Historical Review,* vol. XXIV,

no. 1; *Missouri, the "Show Me" State*, WPA; Frederic A. Culmer, *A New History of Missouri*.
[5] Lewis and Smith, *Chicago, the History of its Reputation*.
[6] Woodward, *History of the St. Louis Bridge*.
[7] Louis How, *James B. Eads*.
[8] Woodward, *History of the St. Louis Bridge*.
[9] Reavis, *St. Louis, the Future Great City*.
[10] Woodward, *History of the St. Louis Bridge*.
[11] Corthell, *James B. Eads* (pamphlet).
[12] Woodward, *History of the St. Louis Bridge*.
[13] *Ibid*.
[14] *Ibid*.
[15] *Ibid*.
[16] Lewis and Smith, *Chicago, the History of Its Reputation*.
[17] Woodward, *History of the St. Louis Bridge;* W. Milnor Roberts, *The Bridge Convention at St. Louis, Aug. 1867*.
[18] James B. Eads, "Review of the Report of the Bridge Convention, St. Louis, Aug. 21, 1867," Woodward, *History of the St. Louis Bridge*.

CHAPTER VIII

[1] (Remark: Woodward mentions both flood levels in some places as "high water," in others he uses the 1828 level as "high water" and the 1844 as "extreme high water." In this book the latter terminology is employed.)
[2] Woodward, *History of the St. Louis Bridge*.
[3] *Ibid*.
[4] *Ibid*.
[5] *Ibid*.
[6] *Ibid*.
[7] *Ibid*.
[8] Snyder, "The Eads of Argyle," *Iowa Journal of History and Politics*, vol. 42, no. 1; J. D. Barnes, "Old Times on the Mississippi," Port Byron (Ill.) *Globe*, in the 1890's (no other date).
[9] Woodward, *History of the St. Louis Bridge*.
[10] *Ibid*.
[11] Louis How, *James B. Eads*.
[12] Woodward, *History of the St. Louis Bridge*.
[13] *Ibid*.
[14] *Ibid*.; David Steinman and Sara Ruth Watson, *Bridges and Their Builders*.
[15] James B. Eads, "Letter to Editor of *Engineering*, London, May 16,

1873," McHenry, *Addresses and Papers;* Eads, "Report to the St. Louis Bridge Company, Sept. 1, 1869," McHenry, *Addresses and Papers;* Eads, "Letter to *Engineering,* London, Nov. 1868," *Addresses and Papers.*
  [16] Woodward, *History of the St. Louis Bridge.*
  [17] *Ibid.;* W. Milnor Roberts, *The Bridge Convention at St. Louis.*
  [18] Woodward, *History of the St. Louis Bridge.*
  [19] Louis How, *James B. Eads.*
  [20] Woodward, *History of the St. Louis Bridge.*
  [21] *Ibid.*
  [22] *Ibid.;* W. Milnor Roberts, *The Bridge Convention at St. Louis.*
  [23] Woodward, *History of the St. Louis Bridge.*
  [24] *Ibid.*
  [25] *Ibid.;* A. Jaminet, M.D., *Physical Effects of Compressed Air.* (Six hundred men worked in the air chambers, fourteen died of caisson disease, according to Woodward.)
  [26] Woodward, *History of the St. Louis Bridge.*
  [27] *Ibid.*
  [28] "James B. Eads," *Central Magazine,* July, 1874.
  [29] Woodward, *History of the St. Louis Bridge;* Jaminet, *Physical Effects of Compressed Air.*
  [30] James B. Eads, "Letter to *Engineering,* London, Sept. 5, 1873," replying to letter of Roebling, McHenry, *Addresses and Papers.*
  [31] *Ibid.*
  [32] Woodward, *History of the St. Louis Bridge.*
  [33] Clipping from St. Louis *Globe-Democrat,* date missing.
  [34] Woodward, *History of the St. Louis Bridge.*
  [35] *Ibid.*
  [36] *Ibid.*
  [37] *Ibid.*
  [38] *Ibid.*

CHAPTER IX

  [1] Woodward, *History of the St. Louis Bridge;* Andrew Carnegie, *Autobiography.*
  [2] (Remark: The theory of the ribbed arch had been worked out by William Chauvenet, Chancellor of Washington University, St. Louis, formerly at Naval Academy, Annapolis. He also devised a method of measuring the elasticity of metal.) (Woodward.)
  [3] Woodward, *History of the St. Louis Bridge:* "The modulus of elasticity is determined by the amount of yielding under a strain which produces no set, and is that strain which bears to the imposed strain the same ratio that the original length of the bar does to its change of length."

NOTES 317

[4] Woodward, *History of the St. Louis Bridge*.
[5] *Ibid.*
[6] *Ibid.*
[7] Snyder, *The St. Louis Movement;* Clem Forbes, "The St. Louis School of Thought," *Missouri Historical Review*, vol. XXIV, no. 1.
[8] Woodward, *History of the St. Louis Bridge*.
[9] James B. Eads, "Address to pupils of night schools of the St. Louis Polytechnic Institute, 1859," McHenry, *Addresses and Papers*.
[10] St. Louis *Missouri Republican*, Jan. 3, 1872, research by Dr. William Swekosky; (Merchants' Exchange was on Main between Walnut and Market, Eads's office at Main and Green); Stevens, *St. Louis, the Fourth City;* "Missouri Entertains a Grand Duke," *Missouri Historical Review*, vol. XXX, no. 4.
[11] Same as note 10.
[12] Same as note 10.
[13] Same as note 10.
[14] St. Louis *Missouri Republican*, Jan. 8 and Jan. 11, 1872.
[15] James B. Eads, "Light," inaugural address before St. Louis Academy of Science, Jan. 15, 1872, McHenry, *Addresses and Papers*.
[16] Woodward, *History of the St. Louis Bridge*.
[17] *Ibid.*
[18] *Ibid.*
[19] U.S. Patent records; *Dictionary of American Biography;* Louis How, *James B. Eads*.
[20] Letter of Eads to Hon. James S. Rollins, U.S. Senate, James Rollins Manuscripts, Missouri Historical Society, Columbia, Mo. Cataloguing by Floyd C. Shoemaker, Secretary.
[21] *Dictionary of American Biography:* Benjamin Gratz Brown.
[22] James B. Eads, letters to *Engineering,* London, McHenry, *Addresses and Papers*.
[23] Woodward, *History of the St. Louis Bridge*.
[24] *Ibid.*
[25] Andrew Carnegie, *Autobiography*.
[26] *Ibid.*
[27] *Ibid.*
[28] *Ibid.*
[29] *Ibid.*
[30] *Ibid.*
[31] *Ibid.;* Corthell, "Address on Retiring from Presidency of the Western Society of Engineers," *Association of Engineering Societies Journal*, vol. IX, no. 5, gave the cost of the Bridge as $6,536,729.99. The $9,000,000 included interest, tunnel, land rights, etc.

[32] Woodward, *History of the St. Louis Bridge.*
[33] *Ibid.;* James B. Eads, *Review of U. S. Engineers' Report on the St. Louis Bridge* (pamphlet).
[34] Woodward, *History of the St. Louis Bridge.*
[35] Ex. Doc. No. 194, H.R., 43rd Congress, 1st session.
[36] Woodward, *History of the St. Louis Bridge.*
[37] *Ibid.*
[38] *Ibid.*
[39] *Ibid.*
[40] Data by Dr. Swekosky; East St. Louis *Journal,* May 26, 1940, Jubilee Edition; Dr. William Taussig, "Personal Recollections of General Grant," *Missouri Historical Society Collections,* vol. II, no. 3; *Fifty Years of Transportation* (brochure), furnished by P. J. Watson, Jr., President of Terminal Association, St. Louis.
[41] Same as note 40.
[42] Same as note 40.
[43] James B. Eads, *Review of U.S. Engineers' Report on the St. Louis Bridge.*
[44] St. Louis *Missouri Republican,* Jan. 18, 1874.
[45] Woodward, *History of the St. Louis Bridge.*

CHAPTER X

[1] Woodward, *History of the St. Louis Bridge.*
[2] St. Louis *Missouri Republican,* Jan. 10, 11 and 18, 1874.
[3] Woodward, *History of the St. Louis Bridge.*
[4] St. Louis *Missouri Republican,* Jan. 15, 1874; St. Louis *Globe-Democrat,* Jan. 15, 1874. Research by Dr. Swekosky.
[5] St. Louis *Missouri Republican,* Jan. 7, 1874.
[6] Woodward, *History of the St. Louis Bridge.*
[7] *Ibid.*
[8] *Ibid.;* Carnegie, *Autobiography.*
[9] Woodward, *History of the St. Louis Bridge;* Carnegie, *Autobiography.*
[10] Woodward, *History of the St. Louis Bridge.*
[11] St. Louis *Missouri Democrat,* June 4, 1874.
[12] St. Louis *Missouri Republican,* July 1, 1874; Woodward, *History of the St. Louis Bridge.*
[13] Scharf, *History of St. Louis.*
[14] *Ibid.*
[15] *Ibid.;* "James B. Eads," *Central Magazine,* July, 1874; St. Louis *Missouri Republican,* July 5, 1874.
[16] Scharf, *History of St. Louis;* Stevens, *St. Louis, the Fourth City;*

Central Magazine, July, 1874; St. Louis *Missouri Republican*, July 4 and 5, 1874.

[17] James B. Eads, "Address at the Opening of the Bridge," McHenry, *Addresses and Papers;* Stevens, *St. Louis, the Fourth City.*

[18] Scharf, *History of St. Louis;* St. Louis *Missouri Republican*, July 5, 1874; *Central Magazine*, July, 1874; Stevens, *St. Louis the Fourth City.*

[19] Snyder, *The St. Louis Movement.*

[20] Gustav Lindenthal, "Bridge Engineering," *Engineering News-Record*, vol. 92, no. 16.

[21] Archibald Henderson, "Eads, Master Engineer," *Universal Engineer*, vol. 55, no. 1.

[22] David B. Steinman and Sara Watson, *Bridges and Their Builders.*

[23] James B. Eads, "Upright Arch Bridges" and "Reply to criticism of this paper," both in McHenry, *Addresses and Papers.*

CHAPTER XI

[1] Rolla Wells, *Episodes of My Life; Fifty Years of Transportation* (brochure), St. Louis Terminal Association.

[2] Reavis, *St. Louis, the Future Great City;* Scharf, *History of St. Louis.*

[3] Elmer L. Corthell, *A History of the Jetties at the Mouth of the Mississippi River.*

[4] *Proceedings of a Congressional Convention on River Improvement, St. Louis, May, 1873* (pamphlet).

[5] Corthell, *History of the Jetties.*

[6] Gould, *Fifty Years on the Mississippi.*

[7] Corthell, *History of the Jetties;* James B. Eads, *Discussion of paper by Colonel Merrill, U.S.A.* (pamphlet).

[8] Corthell, *History of the Jetties;* Reavis, *St. Louis, the Future Great City;* James B. Eads, *Discussion of paper by Colonel Merrill, U.S.A.* (pamphlet).

[9] Corthell, *James B. Eads* (pamphlet).

[10] *Ibid.;* James B. Eads, "Jetty System Explained, Address, Feb. 22, 1874," McHenry, *Addresses and Papers;* Eads, "Letter to Carl Schurz, Jan. 24, 1874," McHenry, *Addresses and Papers;* Eads, *Discussion of paper by Colonel Merrill, U.S.A.*

[11] John Lathrop Matthews, *Remaking the Mississippi;* Eads, *Review of report of U.S.A. Engineers appointed, 1878* (pamphlet); Corthell, *History of the Jetties.*

[12] James B. Eads, "Correspondence with Business Men of New Orleans reviewed, Apr. 6, 1874," McHenry, *Addresses and Papers*, also pamphlet.

[13] Corthell, *James B. Eads* (pamphlet).

[14] Same as note 12.

[15] Corthell, *History of the Jetties.*
[16] *Ibid.;* Rolla Wells, *Episodes of My Life.*
[17] Corthell, *History of the Jetties.*
[18] *Ibid.* (Remark: Southwest Pass was 1,500 feet wide, 40 feet deep, 17 miles long; South Pass was 700 feet wide, 34 feet deep, 12¼ miles long (Corthell).
[19] Corthell, *History of the Jetties.*
[20] New York *Daily Tribune,* May 17, 1878; Eads, "Address at banquet in his honor, St. Louis, March 23, 1875," McHenry, *Addresses and Papers.*
[21] *Ibid.*
[22] *Ibid.*

CHAPTER XII

[1] Corthell, *History of the Jetties;* Corthell, *Ten Years' Practical Teaching in River and Harbor Hydraulics;* Louisiana, *A Guide to the State,* Writers' Project, WPA; New York *Daily Tribune,* March 29, 1879: special correspondent's description of South Pass.
[2] William Starling, "Improvement of the South Pass of the Mississippi River," *Engineering News-Record,* vol. XLIV, no. 1.
[3] Corthell, *History of the Jetties.*
[4] *Ibid.*
[5] *Ibid.*
[6] *Ibid.*
[7] *Ibid.*
[8] *Ibid.*
[9] St. Louis *Globe-Democrat,* May 18 and 19, 1876; Rose Brown, *An American Emperor.*
[10] New Orleans *Daily Picayune,* May 25, 26 and 27, 1876.
[11] New Orleans *Daily Picayune,* May 27, 1876; New Orleans *Times,* May 29, 1876.
[12] Corthell, *James B. Eads* (pamphlet).
[13] Corthell, *History of the Jetties.*
[14] *Ibid.*
[15] *Ibid.*
[16] James B. Eads, "Report to the South Pass Jetty Company, Aug., 1876," McHenry, *Addresses and Papers.*
[17] Edmond Souchon, M.D., *Reminiscences of James B. Eads* (pamphlet).
[18] New Orleans *Times-Democrat,* March 11, 1887.
[19] Corthell, *Ten Years' Practical Teaching in River and Harbor Hydraulics.*
[20] Corthell, *History of the Jetties;* Eads, "Report to the South Pass Jetty Company," McHenry, *Addresses and Papers* (also pamphlet).

[21] Corthell, *History of the Jetties.*
[22] *Ibid.*
[23] James B. Eads, "Letter to Hon. W. S. Holman, Chairman of Committee on Appropriations, Jan. 29, 1877," McHenry, *Addresses and Papers.*
[24] Report No. 632, Senate, 44th Congress, 2nd session; Ex. Doc. No. 28, H.R., 44th Congress, 2nd session.
[25] Corthell, *History of the Jetties.*

CHAPTER XIII

[1] Sylvester Waterhouse, Appendix to *Memorial to Congress for the Improvement of the Mississippi* (pamphlet).
[2] Corthell, *History of the Jetties.*
[3] Albert Warren Kelsey, *Autobiographical Notes and Memoranda;* Scharf, *History of St. Louis;* East St. Louis *Journal,* May 26, 1940.
[4] Same as note 3.
[5] James B. Eads, "Remarks at First Alumni Banquet of Washington University, St. Louis, Feb. 22, 1868," McHenry, *Addresses and Papers.*
[6] Corthell, *History of the Jetties.*
[7] *Ibid.*
[8] *Ibid.*
[9] St. Louis *Missouri Republican,* Oct. 31, 1877, quoting from the Cincinnati *Enquirer.* Research by Dr. Swekosky.
[10] St. Louis *Missouri Republican,* Oct. 31, 1877.
[11] Corthell, *History of the Jetties.*
[12] *Ibid.;* Ex. Doc. No. 37, H.R., 45th Congress, 2nd session.
[13] Corthell, *History of the Jetties.*
[14] "Sketch of James B. Eads," *Popular Science Monthly,* vol. 28; Thomas Frederick Davis, *A History of Jacksonville, Florida.*
[15] Branch Cabell and H. A. Hanna, *The St. Johns, a Parade of Diversities;* Margaret Deland, *Florida Days.*
[16] James B. Eads, "Report of the St. Johns River, March 29, 1878," McHenry, *Addresses and Papers.*
[17] Davis, *A History of Jacksonville, Florida.*
(Remark: The street named for Eads was later named Oak Street.)
[18] Ex. Doc. No. 78, Senate, 45th Congress, 2nd session: Statement of Eads on South Pass jetties; Corthell, *History of the Jetties.*
[19] New York *Daily Tribune,* June 6, 1878; New York *Daily Tribune,* May 17, 1878: letter by Henry Knapp; New York *Daily Tribune,* May 17, 1878, editorial.
[20] Corthell, *History of the Jetties.*
[21] *Ibid.* (Remark: In an address pub. in *Association of Engineering*

Societies Journal, vol. IX, Corthell said the sea-end blocks weighed 181 tons each.)

[22] Corthell, *History of the Jetties.*

[23] Ex. Doc. No. 78, Senate, 45th Congress, 2nd session: Letter of Eads to Secretary of War McCrary; Corthell, *History of the Jetties.*

[24] James B. Eads, *Review of report by Board of U.S. Engineers,* June 19, 1878 (pamphlet).

[25] New York *Daily Tribune,* Apr. 3, 1879.

[26] New York *Daily Tribune,* March 29, 1879.

[27] *Municipal Reference Library Bulletin,* St. Louis, March, 1927; *Fifty Years of Transportation,* Terminal Association, St. Louis; Scharf, *History of St. Louis.*

[28] Same as note 27; Steinman and Watson, *Bridges and Their Builders.* The role of the St. Louis Bridge in the winning of the West was recognized 20 years later by a postage stamp issue picturing the Bridge.

[29] New York *Daily Tribune,* Apr. 3, 1879.

[30] Corthell, *History of the Jetties;* Ex. Doc. No. 49, Senate, 45th Congress, 3rd session.

[31] *Dictionary of American Biography:* James B. Eads; Louis How, *James B. Eads.*

[32] New York *Daily Tribune,* Apr. 3, 1879.

[33] New York *Times,* March 11, 1887; Corthell, *History of the Jetties;* Corthell, *Ten Years' Practical Teaching in River and Harbor Hydraulics* (pamphlet).

[34] New York *Daily Tribune,* editorial, July 15, 1879.

[35] New York *Daily Tribune,* Aug. 11, 1879, quoting Cincinnati *Commercial.*

[36] Gould, *Fifty Years on the Mississippi,* quoting Memphis *Avalanche.*

[37] Corthell, *History of the Jetties.*

[38] "James B. Eads," *Scientific American,* vol. LVI, no. 17.

[39] Harold Sinclair, *Port of New Orleans.*

CHAPTER XIV

[1] "James B. Eads," *Scientific American,* vol. LVI, no. 17; *Dictionary of American Biography:* James B. Eads.

[2] Waterhouse, Appendix to *Memorial to Congress by River Improvement Convention, St. Paul, 1877* (pamphlet).

[3] James B. Eads, "Address at the Grand Convention for Improvement of Mississippi, St. Louis, 1867," McHenry, *Addresses and Papers.*

[4] Benjamin G. Humphreys, *Floods and Levees of the Mississippi.*

[5] *Fortune Magazine,* July, 1942.

⁶ James B. Eads, "Address before Merchants' Exchange, St. Louis, Dec. 5, 1875," McHenry, *Addresses and Papers.*
⁷ Same as note 2.
⁸ Same as note 4.
⁹ Same as note 6.
¹⁰ James B. Eads, "Review of Gen. A. A. Humphreys' letter of May 1, to Hon. E. W. Robertson, H.R.," McHenry, *Addresses and Papers.*
¹¹ James B. Eads, "Review of Humphreys' and Abbott's Report on the Physics and Hydraulics of the Mississippi," McHenry, *Addresses and Papers.*
¹² Same as note 10.
¹³ Rossiter Raymond, "Criticism of James B. Eads's Review of Humphreys' and Abbott's Report on Physics and Hydraulics of the Mississippi," *Engineering and Mining Journal,* Sept. 7, 1886.
¹⁴ "Sketch of James B. Eads," *Popular Science Monthly,* vol. 28.
¹⁵ Corthell, *James B. Eads* (pamphlet).
¹⁶ George Barnett Smith, *Life and Enterprises of Ferdinand de Lesseps.*
¹⁷ Corthell, "Lecture before National Geographic Society, Nov. 22, 1895," printed in Ex. Doc. No. 34, Senate, 54th Congress, 1st session.
¹⁸ *Ibid.;* Ferdinand de Lesseps, "The Oceanic Canal," *North American Review,* January, 1880.
¹⁹ Charles Colné, *The Panama Interoceanic Canal* (pamphlet); Commander Bedford Pim, *The Gate to the Pacific.*
²⁰ Same as note 17.
²¹ Smith, *Life and Enterprises of Ferdinand de Lesseps.*
²² James B. Eads, "Letter to Governor George Perkins, of California, Nov. 8, 1880," McHenry, *Addresses and Papers.*
²³ Sacramento Chamber of Commerce, *The Romance of California* (pamphlet).
²⁴ Letters of James and Martha Eads, loaned by James Eads Switzer.
²⁵ Julian Dana, *The Sacramento, River of Gold.*
²⁶ James B. Eads, "Report to Governor Perkins, of California, Nov. 8, 1880," McHenry, *Addresses and Papers.*
²⁷ *Ibid.*
²⁸ "Ship-Railways, 700 B.C. to 1920 A.D.," *Engineering News-Record,* vol. 85, no. 2; *New International Encyclopedia.*
²⁹ Gould, *Fifty Years on the Mississippi;* Misc. Doc. No. 16, H.R., 46th Congress, 3rd session.
³⁰ James B. Eads, "River Commission Report," Ex. Doc. No. 58, H.R., 46th Congress, 2nd session; Ex. Doc. No. 10, Senate, 47th Congress, 1st session; *Improvement of the Lower Mississippi for Flood Control and Navigation* (3 vols.), prepared under direction of Brigadier General T. H. Jackson and Major D. O. Elliot.

³¹ James B. Eads, "River Commission Report," Ex. Doc. No. 58, H.R., 46th Congress, 2nd session.
³² *Ibid.*
³³ New York *Daily Tribune,* Jan. 22, 1880.
³⁴ New York *Daily Tribune,* Oct. 11, 1880; Ex. Doc. No. 34, Senate, 54th Congress, 1st session.
³⁵ Misc. Doc. No. 16, H.R., 46th Congress, 3rd session.
³⁶ Chauncey D. Griswold, M.D., *The Isthmus of Panama and What I Saw There; Encyclopedia Americana:* Nicaragua.
³⁷ Misc. Doc. No. 16, H.R., 46th Congress, 3rd session.
³⁸ *Ibid.*
³⁹ *Ibid.*
⁴⁰ New York *Daily Tribune,* March 12, 1880.
⁴¹ Misc. Doc. No. 16, H.R., 46th Congress, 3rd session.
⁴² *Ibid.*
⁴³ *Ibid.*
⁴⁴ *Congressional Record,* 46th Congress, 3rd session.
⁴⁵ Corthell, "Address before National Geographic Society," in Ex. Doc. No. 34, Senate, 54th Congress, 1st session.

CHAPTER XV

¹ Mexico City *The Two Republics,* Nov. 28, 1880; *Encyclopaedia Britannica:* Description of Mexico City.
² Mexico City *The Two Republics,* Nov. 28, 1880, and Jan. 30, 1881; "James B. Eads," *Scientific American,* vol. LVI, no. 17.
³ *Encyclopedia Americana:* General Porfirio Diaz.
⁴ Mexico City *El Nacionale,* Apr. 23, 1881, Missouri Historical Society Eads Papers.
⁵ Mexico City *The Two Republics,* Dec. 25, 1880, quoting from New York *Sun* of Nov. 28, 1880.
⁶ *Dictionary of American Biography:* James B. Eads.
⁷ Mexico City *The Two Republics,* Jan. 30, 1881.
⁸ Mexico City *The Two Republics,* Jan. 30, 1881.
⁹ *The Illustrated Scientific News,* vol. I, no. 3.
¹⁰ Mexico City *The Two Republics,* Jan. 30, 1881.
¹¹ Ex. Doc. No. 107, Appendix E, H.R., 47th Congress, 1st session.
¹² Same as note 9.
¹³ Misc. Doc. No. 16, H.R., 46th Congress, 3rd session.
¹⁴ Corthell, *The Atlantic and Pacific Ship-Railway Considered Commercially, Politically and Constructively* (pamphlet).
¹⁵ Data from Wallace McHenry and Dr. William G. Swekosky.

[16] Louis How, *James B. Eads.*
[17] Ex. Doc. No. 55, Senate, 45th Congress, 2nd session, and Ex. Doc. No. 10, Senate, 47th Congress, 1st session.
[18] James B. Eads, "Address before San Francisco Chamber of Commerce, Aug. 11, 1880," McHenry, *Addresses and Papers.*
[19] Louis How, *James B. Eads.*
[20] Misc. Doc. No. 16, H.R., 46th Congress, 3rd session.
[21] *Ibid.;* Report No. 213, Senate, 47th Congress, 1st session.
[22] De Lesseps, "The Panama Canal," *North American Review*, July, 1880.
[23] New York *Times*, July 31, 1879.
[24] Smith, *Life and Enterprises of Ferdinand de Lesseps.*
[25] Mexico City *The Two Republics*, March 27, Apr. 10, and Apr. 17, 1881.
[26] Mexico City *The Two Republics*, Apr. 17, 1881.
[27] Mexico City *El Nacionale*, Apr. 23, 1881, Missouri Historical Society Eads Papers.
[28] Mexico City *The Two Republics*, May 29, 1881.
[29] James B. Eads, "Review of Correspondence with General Grant," New York *Herald*, March 5, 1886.
[30] Mexico City *The Two Republics*, June 19, 1881.
[31] *Ibid.*
[32] James B. Eads, "Report to Hon. H. L. Langevin, Minister of Public Works, Canada, on Harbor at Toronto, 1882," McHenry, *Addresses and Papers* (also pamphlet).
[33] *Dictionary of American Biography:* James B. Eads; William D. Lyman, *The Columbia River;* Robert O. Case, *River of the West; Encyclopedia Americana:* The Columbia River.
[34] Mexico City *The Two Republics*, July 31, 1881 (New York dispatch).
[35] Information from James Eads Switzer, grandson of James B. Eads.
[36] James B. Eads, "Address before British Association for the Advancement of Science, Sept. 5, 1881," McHenry, *Addresses and Papers;* Louis How, *James B. Eads.*
[37] Same as note 36.
[38] Mexico City *The Two Republics*, Oct. 30, 1881.
[39] Mexico City *The Two Republics*, Nov. 24 and 27, and Dec. 15, 1881.
[40] Smith, *Life and Enterprises of Ferdinand de Lesseps.*
[41] U. S. Grant, "The Nicaragua Canal," *North American Review*, February, 1881; Rear Admiral Ammen, *The Errors and Fallacies of the Interoceanic Question—To Whom Do They Belong?* (pamphlet).
[42] *Congressional Record*, 46th Congress, 3rd session.

[43] Mrs. Chapin, *American Court Gossip;* Lewis and Smith, *Oscar Wilde Discovers America, 1882.*
[44] Ex. Doc. No. 107, Appendix E, H.R., 47th Congress, 1st session.
[45] *Ibid.:* embodying letter of James B. Eads to C. G. Williams, Chairman of Committee on Foreign Affairs.
[46] Corthell, *The Atlantic and Pacific Ship-Railway* (pamphlet).
[47] Report No. 213, Senate, 47th Congress, 1st session.
[48] Corthell, *The Atlantic and Pacific Ship-Railway.*
[49] Mrs. Chapin, *American Court Gossip.*
[50] New Orleans *Times-Democrat,* March 18, 1887, quoting New York *Mail and Express.*
[51] New York *Times,* Jan. 29, 1881.
[52] New York *Times,* March 2, 1882.
[53] James B. Eads, "Letter to Joseph Medill," McHenry, *Addresses and Papers.*
[54] *Congressional Record,* 47th Congress, 1st session.
[55] Louis How, *James B. Eads.*
[56] New Orleans *Times-Democrat,* Dec. 7, 1882.

CHAPTER XVI

[1] Corthell, *James B. Eads* (pamphlet).
[2] Corthell, "Address . . . Jan. 8, 1890," *Association of Engineering Societies Journal,* vol. IX, no. 5.
[3] Smith, *Life and Enterprises of Ferdinand de Lesseps.*
[4] Robert S. Taylor, *Improvement of the Mississippi River* (pamphlet).
[5] James B. Eads, "Letter to William Windom, Chairman of Committee on Routes to the Seaboard, March 15, 1874," McHenry, *Addresses and Papers;* "James B. Eads," *Scientific American,* vol. LVI, no. 17; Ex. Doc. No. 10, Senate, 47th Congress, 1st session.
[6] Ex. Doc. No. 10, 47th Congress, 1st session; New York *Daily Tribune,* Apr. 24 and 25, 1882.
[7] New York *Daily Tribune,* Apr. 24 and 25, 1882.
[8] *Ibid.*
[9] *Statement by the Atlantic and Pacific Ship-Railway Company,* June, 1888 (pamphlet).
[10] Data by A. D. Stevens, St. Louis.
[11] Same as note 9.
[12] New Orleans *Times-Democrat,* 1884 (date missing); "Remarks by J. G. Chapman," *Proceedings of Public Meeting at Southern Hotel, St. Louis, Nov. 4, 1885* (pamphlet).
[13] *Proceedings of Convention for the Improvement of the Mississippi and Tributaries, Washington, D. C., February, 1884.*

NOTES 327

[14] Mrs. Chapin, *American Court Gossip.*
[15] Same as note 13.
[16] *Congressional Record*, 48th Congress, 1st session.

CHAPTER XVII

[1] "James B. Eads," *Scientific American*, vol. LVI, no. 17; Corthell, "Address, Jan. 8, 1890," *Association of Engineering Societies Journal*, vol. IX, no. 5. (Remark: In July, 1886, contract for building the canal was let for $28,750,000.)
[2] *Ibid.*
[3] *Engineering*, London, June 13, 1884, vol. 37.
[4] *Engineering*, London, July 18, 1884, vol. 38.
[5] New York County records, furnished by First Deputy Register Lewis Orgel; description from viewing the house, still standing.
[6] *Valentine's Manual of New York*, edited by Henry Collins Brown.
[7] *Ibid.*
[8] *Ibid.*
[9] Letter from a relative of Martha Eads to author.
[10] Louis How, *James B. Eads; Dictionary of American Biography.*
[11] St. Louis *Missouri Republican*, May 10, 1885; Louis How, *James B. Eads.*
[12] Louis How, *James B. Eads.*
[13] Davis, *History of Jacksonville, Florida; Encyclopedia Americana:* St. Johns River, and Jacksonville, Florida.
[14] James B. Eads, *Discussion of paper by Colonel Merrill* (pamphlet).
[15] J. O. Dyer, *The Early History of Galveston;* Jesse A. Ziegler, *Wave of the Gulf.*
[16] James B. Eads, *Discussion of paper by Colonel Merrill* (pamphlet).
[17] *Ibid.*
[18] New York *Times*, March 13, 1885, letter of Eads to editor.
[19] James B. Eads, *Discussion of paper by Colonel Merrill* (pamphlet).
[20] *Ibid.*
[21] *Ibid.*
[22] New York *Times*, March 13, 1885.
[23] Same as note 19.
[24] *Speeches in the 48th Congress*, 2nd session.
[25] *Ibid.*
[26] Same as note 19.
[27] New York *Times*, March 13, 1885.
[28] Corthell, *Ten Years' Practical Teaching in River and Harbor Hydraulics.*

[29] Remark: Galveston jetties were again being constructed, successfully, Captain O. A. Ernst, U.S.A., in charge.
[30] New Orleans *Times-Democrat*, 1884 (date missing).
[31] New York *Tribune*, Nov. 13, 1884.

CHAPTER XVIII

[1] Smith, *Life and Enterprises of Ferdinand de Lesseps*.
[2] *Proceedings of Public Meeting, Southern Hotel, St. Louis, Nov. 4, 1885* (pamphlet).
[3] *Ibid*.
[4] Smith, *Life and Enterprises of Ferdinand de Lesseps*.
[5] *Ibid*.
[6] Allan Nevins, *Grover Cleveland, A Study in Courage*.
[7] Report No. 717, H.R., 49th Congress, 1st session, including James B. Eads's report to Secretary of State Bayard, May 29, 1885.
[8] Corthell, *Statement before Subcommittee, H.R., Feb. 5, 1886* (pamphlet).
[9] Rear Admiral Ammen, *The Certainty of the Nicaragua Canal* (pamphlet); Corthell, *An Exposition of the Errors and Fallacies in Rear Admiral Ammen's "The Certainty of the Nicaragua Canal"* (pamphlet).
[10] Ammen, *The Errors and Fallacies of the Interoceanic Question—To Whom Do They Belong?* (pamphlet).
[11] *Ibid*.
[12] New York *Herald*, March 6, 1886.
[13] *Encyclopaedia Britannica:* Grover Cleveland.
[14] *Engineering, London*, Dec. 25, 1885.
[15] New York *Daily Tribune*, March 11, 1887.
[16] *Ibid*.
[17] Nevins, *Grover Cleveland, A Study in Courage*.
[18] Ex. Doc. No. 130, Senate, 47th Congress, 1st session; Ex. Doc. No. 56, Senate, 47th Congress, 1st session.
[19] Anecdote by James Eads Switzer, grandson of James B. Eads.
[20] Letter of James B. Eads to Mr. Hagedorn, New York Historical Society.
[21] New York *Times*, Dec. 6, 1886.
[22] New Orleans *Times-Democrat*, March 15, 1887.
[23] New York *Daily Tribune*, March 8, 1886: letter from Eads; New York *Times*, Dec. 6, 1886, editorial.
[24] New York *Daily Tribune*, March 8, 1886: letter from Eads.
[25] Corthell, *James B. Eads* (pamphlet).
[26] James B. Eads, "Review of Correspondence with General Grant," New York *Herald*, March 5, 1886.

NOTES 329

[27] St. Louis *Missouri Republican,* March 11, 1887.
[28] Corthell, *James B. Eads* (pamphlet).
[29] Ex. Doc. No. 34, Senate, 54th Congress, 1st session; U.S. Patent Office records.
[30] New Orleans *Times-Democrat,* March 15, 1887.
[31] New York *Daily Tribune,* March 11, 1887.
[32] New Orleans *Times-Democrat,* March 21, 1887.
[33] Corthell, *James B. Eads* (pamphlet).
[34] *Ibid.*

CHAPTER XIX

[1] New Orleans *Times-Democrat,* March 11, 1887, quoting the Augusta *Chronicle.*
[2] New Orleans *Times-Democrat,* March 18, 1887.
[3] *Ibid.*
[4] St. Louis *Missouri Republican,* March 18, 1887.
[5] *Ibid.*
[6] St. Louis *Missouri Republican,* March 20, 1887.
[7] *Organization and By-laws of the Atlantic and Pacific Ship-Railway Company, 1888,* and *Statement by the Atlantic and Pacific Ship-Railway Company, June, 1888* (pamphlets).
[8] Same as note 7.
[9] Smith, *Life and Enterprises of Ferdinand de Lesseps.*
[10] *Ibid.*
[11] Gould, *Fifty Years on the Mississipi.*
[12] Corthell, "Address before National Geographic Society," printed in Ex. Doc. No. 34, Senate, 54th Congress, 1st session.
[13] *Encyclopedia Americana:* Panama Canal.
[14] "Ship-Railways, 700 B.C. to 1920 A.D.," *Engineering News-Record,* vol. 85, no. 2.
[15] New York *Times,* Aug. 19, 1942.
[16] *Encyclopaedia Britannica:* Floating Docks.
[17] Allen P. Armagnac, "Enlarging the Panama Canal for Bigger Battleships," *Popular Science Monthly,* September, 1940.
[18] Virginia Lee Warren, "Ship Transport by Rail Proposed," New York *Times,* Aug. 22, 1946.
[19] Irving Dilliard, "Eads Bridge, at 72, Having Face Lifted," St. Louis *Post-Dispatch,* Oct. 23, 1946.
[20] St. Louis *Globe-Democrat,* March 16 and 18, 1947.
[21] Corthell, *Ten Years' Practical Teaching in River and Harbor Hydraulics;* Allen E. Washburn, *An Open Mouth for the Mississippi River* (pamphlet).

[22] New Orleans *Times-Picayune,* Oct. 27 and 28, 1929.

[23] *Fortune Magazine,* July, 1942.

[24] *Waterways Journal,* St. Louis, Feb. 26, 1944: address of Ernst Holzborn; also, editorial.

[25] James B. Eads, "Address to the Pupils of Night Schools of the Polytechnic Institute, St. Louis, 1859," McHenry, *Addresses and Papers.*

[26] *Engineering News-Record,* vol. 92, nos. 19 and 20; *Missouri Historical Review,* vol. XXXIII, no. 3.

CONDENSED BIBLIOGRAPHY

AMMEN, DANIEL, Rear-Admiral U.S.N., *Interoceanic Ship Canal Across the American Isthmus*, Bulletin of the American Geographic Society, New York, 1878.
———, "Mr. de Lesseps and His Canal," *North American Review*, February, 1880.
———, "The Nicaragua Route to the Pacific," *North American Review*, November, 1880.
———, *The Errors and Fallacies of the Interoceanic Question—To Whom Do They Belong?* (pamphlet), New York, 1886.
———, *The Certainty of the Nicaragua Canal Contrasted with the Uncertainties of the Eads Ship-Railway* (pamphlet), Washington, 1886.
ANDERSON, GALUSHA, *The Story of a Border City During the Civil War*, Boston, 1908.
ARMAGNAC, ALDEN P., "Enlarging the Panama Canal for Bigger Battleships," *Popular Science Monthly*, September, 1940.
BAIRD, ROBERT, *View of the Valley of the Mississippi*, Philadelphia, 1832.
BATCHELLER, OLIVER A., "The Battle of Mobile Bay," *Magazine of History*, vol. XIV, no. 6.
Battles and Leaders of the Civil War (4 vols.), edited by Robert Underwood Johnson and Clarence Buel, New York, 1884-1887.
BOYNTON, CHARLES BRANDON, D.D., *History of the U.S. Navy During the Rebellion* (2 vols.), New York, 1867.
BROCKETT, LINUS PIERPONT, *Our Great Captains*, New York, 1865.
BROWN, HENRY COLLINS, *Valentine's Manual of Old New York*, U.S.A., 1926.
BROWN, ROSE, *An American Emperor*, New York, 1945.
CABELL, BRANCH, and H. A. HANNA, *The St. Johns, a Parade of Diversities*, New York, 1943.
CARNEGIE, ANDREW, *Autobiography*, Boston, 1920.
CARTER, HODDING, *The Lower Mississippi*, New York, 1942.
CASE, ROBERT ORMOND, *River of the West*, Portland, Ore., 1940.
Central Magazine, St. Louis, July, 1874.
CHAPIN, MRS. (ELIZABETH), *American Court Gossip, or Life at the National Capital*, Marshalltown, Iowa, 1887.

Colné, Charles, *The Panama Interoceanic Canal* (pamphlet), 1884. (Paper read at Franklin Institute, Philadelphia, Oct. 22, 1884.)

Corthell, Elmer Lawrence, D.S., C.E., *A History of the Jetties at the Mouth of the Mississippi River* (2nd edition), New York, 1881.

———, *Ten Years' Practical Teaching in River and Harbor Hydraulics* (pamphlet). (Paper read before the North American Society of Civil Engineers, June 10, 1884.)

———, *The Tehuantepec Ship-Railway* (pamphlet). (Address at Franklin Institute, Dec. 28, 1884.)

———, *The Interoceanic Problem and Its Scientific Solution* (pamphlet). (Address before American Association for the Advancement of Science, Ann Arbor, Mich., Aug. 26, 1885.)

———, *The Atlantic and Pacific Ship-Railway Considered Commercially, Politically and Constuctively* (pamphlet), 1886.

———, *Statement to Subcommittee of Committee on Commerce, H.R., February 5, 1886* (pamphlet).

———, *An Exposition of the Errors and Fallacies in Rear-Admiral Ammen's Pamphlet entitled: "The Certainty of the Nicaragua Canal"* (pamphlet), April, 1886.

———, *Address at the American Shipping Convention, Pensacola, Florida, November 16, 1886* (pamphlet).

———, "Address on Retiring from the Presidency of the Western Society of Engineers, January 8, 1890," *Association of Engineering Societies Journal*, vol. IX, no. 5.

———, *James B. Eads* (pamphlet). (Address to Western Society of Engineers, June 4, 1890.)

———, "The Tehuantepec Route," in Ex. Doc. No. 34, Senate, 54th Congress, 1st session. (Lecture before the National Geographic Society, Washington, Nov. 22, 1895.)

Croly, David G., *Seymour and Blair*, New York, 1868.

Culmer, Frederic Arthur, *A New History of Missouri*, Mexico, Mo., 1938.

Dana, Julian, *The Sacramento, River of Gold*, New York, 1939.

Davis, Thomas Frederick, *A History of Jacksonville, Florida, and Vicinity*, Florida, 1925.

Dayton, Fred Erving, *Steamboat Days*, New York, 1925.

Deland, Margaret, *Florida Days*, Boston, 1889.

Dictionary of American Biography, New York, 1930.

DIXON, FRANK HAIGH, "A Traffic History of the Mississippi System," *National Waterways Commission*, Doc. 11, December, 1909.

DORSEY, DOROTHY B., "The Panic and Depression of 1837-1843 in Missouri," *Missouri Historical Review*, vol. XXX, no. 2.

DUDENS, GOTTFRIED, "Report, 1824-1827," translated by William Bek, *Missouri Historical Review*, vol. XXX, no. 2.

DYER, DR. J. O., *The Early History of Galveston*, Galveston, 1916.

EADS, JAMES B., Addresses, letters and reports, 1859-1884, compiled by Estill McHenry, *Addresses and Papers of James B. Eads*, St. Louis, 1884.

———, Letters to Martha Nash Dillon Eads, loaned by James Eads Switzer, grandson of James B. Eads.

———, *A System of Naval Defences*, New York, 1868. (Report to the Navy Department, Feb. 22, 1868.)

———, *Inaugural Address before the St. Louis Academy of Science*, Jan. 15, 1872 (pamphlet).

———, "Recollections of Foote and the Gunboats," *Battles and Leaders of the Civil War* (4 vols.), vol. I.

———, *Review of the U.S.A. Engineers' Report on the St. Louis Bridge* (pamphlet), St. Louis, 1873.

———, *Report on the Physics and Hydraulics of the Mississippi River to Hon. S. A. Hurlburt, H.R., May 29, 1874* (pamphlet).

———, *Review of Report by Humphreys and Abbott on the Physics and Hydraulics of the Mississippi River* (pamphlet), 1878.

———, *Review of Report by Board of U.S.A. Engineers on the Mississippi Jetties, 1878* (pamphlet).

———, *Address before House of Representatives Committee on Interoceanic Routes, March 9, 1880* (pamphlet). (From Misc. Doc., No. 16, H.R., 46th Congress, 3rd session.)

———, *Interoceanic Ship-Railway* (pamphlet). (Address before San Francisco Chamber of Commerce, Aug. 11, 1880.)

———, "The Isthmian Ship-Railway," *North American Review*, March, 1881.

———, *Provisions of the Ship-Railway Concession from the Mexican Republic* (pamphlet). (Report by Eads to Secretary of State Bayard, May 29, 1885.)

———, *South Pass Jetties: Discussion of paper by Colonel Merrill, U.S.A.* (pamphlet), 1885. (Both Eads and Merrill's papers were read before the American Society of Engineers.)

———, *Review of Correspondence with General Grant* (pamphlet), 1886. (Earlier in the New York *Herald*, March 5, 1886.)

Eads Papers, letters to James B. Eads and other papers, at the Missouri Historical Society, St. Louis, Mo.

EDWARDS, RICHARD, and M. HOPEWELL, *The Great West and Her Commercial Metropolis, St. Louis*, St. Louis, 1860.

ELLET, ALFRED W., "Ellet and His Steam Rams at Memphis," *Battles and Leaders of the Civil War*, vol. I.

Encyclopedia Americana.

Encyclopaedia Britannica.

Engineer, The, Sept. 1, 1939.

Engineering, London, vols. 37, 39 and 40.

Engineering News-Record, vol. 85, no. 2; vol. 92, nos. 19 and 20.

Executive and Miscellaneous Documents of the U.S. Congress, as indicated in notes.

FARRAGUT, LOYALL, *Life of David Glasgow Farragut, First Admiral of the U.S. Navy,* New York, 1882.

Final Proceedings of Advisory Commission of Engineers Convened by James B. Eads to Consider His Plans for the Improvement of the Mouth of the Mississippi, November 18, 1875 (pamphlet).

Fortune Magazine: "Wartime and the River," July, 1942.

GOULD, EMERSON, *Fifty Years on the Mississippi,* St. Louis, 1889.

GRANT, ULYSSES S., "The Nicaragua Canal," *North American Review,* February, 1881.

———, *Personal Memoirs of U. S. Grant* (2 vols.), New York, 1885.

GRISWOLD, CHAUNCEY D., M.D., *The Isthmus of Panama and What I Saw There,* New York, 1852.

HANSON, JOSEPH MILLS, *The Conquest of the Missouri,* Chicago, 1909.

HAZARD, ROWLAND GIBSON, *The Financial and Political Condition of the United States* (pamphlet), London, 1864.

HEADLEY, J. T., *Farragut and Our Naval Commanders,* New York, 1867.

HENDERSON, ARCHIBALD, Ph.D., D.C.L., LL.D., "Eads, Master Engineer," *Universal Engineer,* vol. LV, no. 1 (January, 1932).

HOPPIN, JAMES MASON, *Life of Andrew Hull Foote, Rear-Admiral of the U.S. Navy,* New York, 1874.

How, LOUIS, *James B. Eads,* Boston, 1910.

HUMPHREYS, GEN. A. A. and ABBOTT, *Report on the Physics and Hydraulics of the Mississippi River,* Washington, 1867.

Illustrated Scientific News: "The Eads Ship-Railway," vol. I, no. 3.
INMAN, HENRY, *The Old Santa Fe Trail,* New York, 1897.
International Year Book, 1941: "Improvement of the Mississippi River," New York, 1941.
Iowa Journal of History and Politics: "The Eads of Argyle," vol. 42, no. 1.
JAMINET, A., M.D., *Physical Effects of Compressed Air* (pamphlet), St. Louis, 1871.
JOHNSON, ROBERT UNDERWOOD, and CLARENCE BUEL, *Battles and Leaders of the Civil War* (4 vols.), New York, 1884-1887.
KELSEY, ALBERT WARREN, *Autobiographical Notes and Memoranda,* Baltimore, 1911.
KING, J. W., U.S.N., *Report to the Navy Department of the Eads Steam Turret, April 30, 1864* (pamphlet), Washington.
KNOX, THOMAS, *Robert Fulton,* New York, 1886.
LESSEPS, FERDINAND DE, "The Interoceanic Canal," *North American Review,* January, 1880.
———, "The Panama Canal," *North American Review,* July, 1880.
LEWIS, LLOYD, and HENRY JUSTIN SMITH, *Chicago: The History of Its Reputation,* New York, 1929.
LINDENTHAL, GUSTAV, "Bridge Engineering," *Engineering News-Record,* vol. 92, no. 16.
LYMAN, WILLIAM DENISON, *The Columbia River,* New York, 1917.
Magazine of History, vol. LV, no. 6.
MAHAN, CAPTAIN A. T., U.S.N., *The Navy in the Civil War—The Gulf and Inland Waters,* New York, 1883.
Marine Journal: "Strong Endorsement of the Ship-Railway," Nov. 20, 1886.
MATTHEWS, JOHN LATHROP, *Remaking the Mississippi,* Boston, 1909.
MCCLELLAN, ELISABETH, *Historic Dress in America, 1800-1870,* Philadelphia, 1910.
MCHENRY, ESTILL, *Addresses and Papers of James B. Eads,* St. Louis, 1884.
MERRICK, GEORGE BYRON, *Old Times on the Upper Mississippi,* Cleveland, 1909.
MILLER, FRANCIS TREVELYAN, *The Photographic History of the Civil War* (10 vols.), New York, 1912.
Missouri Historical Review, Columbia, Mo., as indicated in notes.
Missouri Historical Society Collections, vol. II, no. 3, St. Louis.

Missouri Historical Society Eads Papers, St. Louis, as indicated in notes.
Municipal Reference Library Bulletin, St. Louis, March, 1927.
New International Encyclopedia: Ship-railway.
Official Records of the Union and Confederate Armies in the War of the Rebellion, Washington, 1800-1901, vols. as indicated in notes.
Official Records of the Union and Confederate Navies in the War of the Rebellion, Washington, 1894-1922, vols. as indicated in notes.
Organization and By-laws of the Atlantic and Pacific Ship-Railway Company, 1888 (pamphlet).
PARKER, COMMODORE FOXALL, U.S.N., *The Battle of Mobile Bay,* Boston, 1878.
PHELPS, ALBERT, *Louisiana—A Record of Expansion,* Boston, 1905.
PIM, COMMANDER BEDFORD, R.N., *Gate of the Pacific,* Boston, 1863.
Popular Science Monthly: "Sketch of James B. Eads," vol. 28.
Proceedings of Bridge Convention, St. Louis, August, 1867 (pamphlet).
REAVIS, LOGAN U., *Saint Louis, the Future Great City of the World.* St. Louis, 1875.
ROBERTS, W. MILNOR, *The Bridge Convention at St. Louis, August, 1867* (pamphlet).
ROLLINS, CURTIS B., "Some Impressions of Frank P. Blair," *Missouri Historical Review,* vol. XXIV, no. 3.
Sacramento Chamber of Commerce, *The Romance of California* (pamphlet).
SCHARF, JOHN THOMAS, *A History of St. Louis, City and County,* Philadelphia, 1883.
Scientific American: "James B. Eads," vol. LVI, no. 17.
SELLERS, WILLIAM, *Memoir of James B. Eads* (pamphlet), 1895. (Read before the National Academy, 1888.)
SHOEMAKER, FLOYD CALVIN, *Missouri's Hall of Fame,* Columbia, Mo., 1923.
———, *Missouri and Missourians, Land of Contrasts and People of Achievements* (5 vols.), Chicago, 1943.
SHOEMAKER, FLOYD CALVIN, and WALTER WILLIAMS, *Missouri, Mother of the West* (5 vols.), Chicago and New York, 1930.
SINCLAIR, HAROLD, *Port of New Orleans,* New York, 1942.
SMITH, GEORGE BARNETT, *The Life and Enterprises of Ferdinand de Lesseps,* London, 1895.

SMITH, GEORGE WASHINGTON, *A History of Illinois,* Chicago and New York, 1927.
SMITH, HENRY JUSTIN, *The War with Mexico,* New York, 1919.
SMITH, JOHN KENDALL, *History of New Orleans,* Chicago and New York, 1922.
SMITH, WILLIAM ERNEST, *The Francis Preston Blair Family in Politics* (2 vols.), New York, 1933.
SNEAD, THOMAS L., *The Fight for Missouri,* New York, 1886.
SNYDER, CHARLES E., "The Eads of Argyle," *Iowa Journal of History and Politics,* vol. 42, no. 1.
SNYDER, DENTON JAQUES, *The St. Louis Movement,* St. Louis, 1920.
SOUCHON, EDWARD, M.D., *Reminiscences of James B. Eads* (pamphlet). (Read before the Louisiana Historical Society, May 19, 1915.)
SPENCER, THOMAS EDWIN, *The Story of Old St. Louis,* St. Louis, 1914.
St. Louis Terminal Company, *Fifty Years of Transportation* (brochure), St. Louis, 1944.
STARLING, WILLIAM, "Improvement of South Pass of the Mississippi River," *Engineering News,* vol. XLIV, no. 6.
Statement by the Atlantic and Pacific Ship-Railway Company (pamphlet), 1888.
STEINMAN, DAVID B., and SARA RUTH WATSON, *Bridges and Their Builders,* New York, 1941.
STEVENS, WALTER BARLOW, "Lincoln in Missouri," *Missouri Historical Review,* vol. X, no. 2.
———, "The Political Turmoil in Missouri in 1874," *Missouri Historical Review,* vol. XXXI, no. 1.
———, *Saint Louis, the Fourth City,* St. Louis and Chicago, 1911.
TAUSSIG, WILLIAM, "Personal Recollections of General Grant," *Missouri Historical Society Collections,* vol. II, no. 3, St. Louis.
TAYLOR, J. N., and M. O. CROOKS, *Sketch Book of St. Louis,* St. Louis, 1858.
TAYLOR, ROBERT S., *The Improvement of the Mississippi River* (address in St. Louis, Jan. 26, 1884).
WARREN, BREVET-GENERAL GOUVERNEUR K., "Report on Bridging the Mississippi," Appendix 3, Annual Report of Chief of Engineers, (pamphlet), Washington, 1878.
WASHBURN, ALLEN E., *An Open Mouth for the Mississippi River* (pamphlet). (Read before Louisiana Engineering Societies, Dec. 8, 1919.)

WATERHOUSE, SYLVESTER, Appendix to *Memorial to Congress for Improvement of the Mississippi, Convention, St. Paul, 1877* (pamphlet).
Waterways Journal, St. Louis, Mo., as indicated in notes.
WELLS, ROLLA, *Episodes of My Life*, St. Louis, 1933.
WILLIAMS, HELEN DAVAULT, "Social Life in St. Louis, 1840-60," *Missouri Historical Review*, vol. XXXI, no. 1.
WOODWARD, CALVIN MILTON, Dean of the Polytechnic School, Washington University, St. Louis, *A History of the St. Louis Bridge*, St. Louis, 1881.
Writers' Program, WPA: *Louisiana—A Guide to the State*, New York, 1941.
Writers' Program, WPA: *Missouri, the "Show-Me" State*, New York, 1941.
ZIEGLER, JESSE A., *Wave of the Gulf*, San Antonio, Texas, 1938.
ZIMMERMAN, EDUARD, "Travels in Missouri in October, 1838," *Missouri Historical Review*, vol. IX, no. 1.

www.ingramcontent.com/pod-product-compliance
Lightning Source LLC
Chambersburg PA
CBHW021817300426
44114CB00009BA/208